ALSO BY HANS CHRISTIAN VON BAEYER

Rainbows, Snowflakes, and Quarks

Taming the Atom

The Fermi Solution

WARMTH DISPERSES AND TIME PASSES

WARMTH DISPERSES AND TIME PASSES

The History of Heat

HANS CHRISTIAN VON BAEYER

Previously published as
Maxwell's Demon

THE MODERN LIBRARY
NEW YORK

1999 Modern Library Paperback Edition

 Published in the United States by Random House, Inc., New York, and simultaneously in Canada by Random House of Canada Limited, Toronto.

The work was originally published in hardcover by Random House, Inc., in 1998, as *Maxwell's Demon*.

ISBN: 0-375-75372-9

Modern Library website address: www.modernlibrary.com

Printed in the United States of America
2 4 6 8 9 7 5 3

To the memory of Rolf Winter (1928–1992),

and to all my other colleagues in the Physics Department

of the College of William and Mary—

for thirty years of friendship and support.

CONTENTS

INTRODUCTION

Imagination is more important than knowledge.

—ALBERT EINSTEIN

Just as all the world's literature revolves ultimately about the simple, inexhaustible themes of love and death, physics begins and ends with the mundane miracles we witness in our daily lives. Why does an apple fall from its branch? With what unseen tentacles does the ground reach up to snare it? Six-year-old children and physicists with million-dollar instruments are equally enthralled by the enigma of gravity. What determines the exquisite symmetry of a snowflake? What gossamer stuff is light made of? Such questions are as universal as curiosity, as ancient as philosophy, as fundamental as language; the answers offered up by philosophers and scientists since the beginning of civilization help to invest our universe with meaning.

Among the multitude of natural phenomena our senses experience and our minds ponder, the feeling of warmth stands out as the most primitive. It is impossible to imagine life without warmth. All organisms require heat—however little it may be—for survival. One theory of the origin of life holds that once upon a time, three or four billion years ago, lightning struck a puddle of water containing a kind of warm chemical chicken broth and triggered the formation of amino acids, which are the building blocks of life. Warmth played the same role in that primal scene that it plays today in the womb. It sustains life, and its absence is tantamount to death.

Long after the emergence of life, it was the control over that strange but indispensable commodity of heat that signaled the birth of civilization. When our ancestors learned to manipulate fire, they crossed the threshold from animal to human, and started off on the long journey toward those simple questions that continue to haunt us: What is warmth? How is it created? Into what hidden crevices does it escape when we no longer feel it?

Today we know that warmth, or heat, is nothing but motion—a palpable manifestation of the ceaseless, random, chaotic agitation of the invisible atoms and molecules that compose all matter. Temperature, a measure of the intensity of heat, turns out to be related to the speed of those particles—the faster they jiggle, the hotter they seem to the touch. These insights in turn suggest new, more difficult questions.

Imagine a hot mug of tea left to cool on the kitchen counter. Slowly and silently, invisible warmth oozes out of the cup into the air. The agitation of the water molecules gradually diminishes while that of the surrounding air molecules imperceptibly increases. But why must it be that way? Couldn't the air molecules bumping into the tea at its surface give up some of their motion—like bowling balls crashing into pins? Why doesn't a warm cup ever heat up spontaneously, and come to a rolling boil, while the vast reservoir of motion vested in the air of even a cool kitchen is imperceptibly reduced?

In the middle of the nineteenth century, this question occupied center stage of scientific research. In the language of thermodynamics, the science of heat that was then emerging, the question became: When a hot body is in contact with a cool body, why does heat always flow from hot to cold—never the other way around? Why does heat only flow downhill, as it were? Phrased in this way, the question recalls the older question—why does the apple fall?—which, in time, led to the theories of universal gravitation and general relativity, and today inspires the search for a quantum theory of gravity. The icons of the falling apple and the cooling mug of tea represent themes as powerful in their domain as love and death are in theirs.

As the Victorian physicists pondered the mystery of the teacup, they discovered that it is closely related to other questions with much graver consequences. Among the most significant of these was the problem of the pathetic inefficiency of steam engines. It also lies at the heart of more modern troubles, for example the unfortunate fact that cars are inherently very inefficient. Better design can improve them, but the maximum efficiency of the most state-of-the-art gasoline engine is approximately 30 percent. This distressing figure implies that more than two thirds of the gas used up is wasted right at the outset—before inadequate design and shoddy tires even begin to take their toll. More than two thirds of the gas we buy goes directly into heating the air through which we drive—and there is *absolutely nothing we can do about it*.

Now, why is that? Why can't a clever engineer come along and capture the heat that escapes from the radiator and the exhaust pipe, turn it around, convert it back into driving power, and save two thirds of the gas we waste? The answer, astonishingly, is that what frustrates such a scheme is the same universal law of nature that prevents a mug of tea from coming to a boil on its own.

Seen in this light, understanding the seemingly innocuous problem of the teacup emerges as an intellectual challenge of considerable urgency. Saving two thirds of the fuel wasted by heat engines would eliminate America's dependence on oil imports, reduce the danger of global warming due to air pollution by smokestack emissions, and dramatically improve the economy of the energy-starved Third World. But it can't be done.

Aside from its troubling practical consequences, the puzzle of the teacup also affects our lives in a more personal and philosophical way. It turns out that the flow of heat from the tea into the air—or from any body into a cooler one, for that matter—defines the flow of time itself! "Earlier" is, by definition, the time when the cup was hot, "later" the time when it is cool. Without this "arrow of time" provided by thermodynamics, we would be living in a reversible world, in which time could flow forward, backward, or not at all. It is difficult to imagine such a world, but a simple example serves to illustrate it. Consider a set of billiard balls rolling around on a table, colliding with each other and the side cushions without friction, air resistance, or even clicking sounds to sap them of their energy. A movie of this scene would hold no clue for a viewer to decide whether the film is running forward or backward: It would be absolutely reversible. Real life and real billiards are not like that. We sense the inexorable forward march of time, even if we cannot put it into adequate words. Ultimately that march is measured by the flow of heat, which is in turn related to the inefficiency of cars.

In their efforts to parse the problem of the teacup, nineteenth-century physicists had at their disposal a powerful new tool: the principle of energy conservation.

Energy—defined most simply as the ability to make something move—comes in many guises. Energy of motion, or kinetic energy, resides in all moving bodies, because they can cause other objects to move by simply knocking into them. Since heat is motion, it too is a type of energy. Other kinds of energy can drive motors, which in turn move things. Thus chemical energy is stored in gasoline and coal, electrical energy courses through metallic wires, radiant energy pours down on us from the sun, and nuclear energy fuels power plants as well as the sun. But energy can also hide in more subtle places. A quiet lake on top of a mountain, for example, moves nothing, powers no motors, and does no work. But if the water is released and runs downhill through a pipe, it can run a waterwheel in the valley. So even the placid lake, appearances to the contrary, stores something called "gravitational potential energy," or simply potential energy. Today we recognize dozens of forms of energy flowing through every cranny and resting in every nook of the universe.

Around 1850, several scientists simultaneously discovered that energy—an abstract, intangible, invisible, and ultimately unimaginable quantity—is conserved. It cannot be created out of nothing, and it cannot be destroyed.

The concept of energy conservation was one of the boldest and most comprehensive scientific syntheses of the nineteenth century, and remains part of the solid, irreducible bedrock of modern physics. When the science of heat, or thermodynamics, was formally established, it was obvious that energy conservation would have to play a leading role in it. For this reason, the law of conservation of energy, as applied specifically to the form of energy called heat, was enshrined as the *first law of thermodynamics.*

But the problem of the teacup, with its implications about the inefficiency of engines and the arrow of time, remained. Why does heat only flow downhill? If you place a hot potato on a cool plate, there is nothing in the law of conservation of energy to keep the plate from giving up more of its meager store of heat to the potato, thus making it even hotter: Nothing prevents the rich getting richer and the poor poorer. So the proposition that "heat flows naturally from hot objects into cooler objects, never the other way around," since it couldn't be explained, was simply enshrined as the *second law of thermodynamics.* But no one understood its origin, or its universality. Does it hold everywhere, without exception, like the first law? Or are there ways of getting around it? Those were the cardinal questions of physics a century and a half ago; they are the subject of this book.

Frustrated by these mysteries, James Clerk Maxwell, the Scottish physicist who created the theory of electromagnetism, and in the process discovered the true nature of light, dreamed up an imaginary being designed to help him test the laws of thermodynamics. It began life as his own private creature, like a witch's familiar, but soon escaped and began to haunt the house of science. Occasionally you glimpse this goblin as it tiptoes across lacy laser beams lined up on an optical table, or peers out from under a computer console, but most of the time it hides behind some massive monograph in the library. Sometimes it gets caught, and unceremoniously thrown out the nearest window—but inevitably it manages to pop back in through another one. Thus it has teased physicists for more than a century and a quarter—an indomitable, insolent gadfly on the warpath against intellectual complacency. They don't know how to cope with it, so they usually just try to ignore it—but it refuses to go away. Its name is Maxwell's Demon.

Maxwell didn't bother to describe the Demon's outward appearance when he invented him, so I feel free to fill in the blank. I think of the Demon as a puckish little elf from an illustration of a Victorian fairy tale. Except for a tattered black loincloth he is naked. His smooth leathery skin glistens like polished mahogany. His limbs are thin, but tough and wiry—the fingers slim, flexible, and extraordinarily deft, the long toes prehensile. His round, slightly bulging eyes regard the world with an unblinking gaze from deep sockets above a noble, aquiline nose. The startling whiteness of his bushy hair and neatly trimmed Vandyke beard contrasts sharply with the dark skin, and provides the only hint of his prodigious age. Sparks of intelligence crackle from his tiny wrinkled face.

The purpose of the Demon was to help Maxwell explore the origins of the second law in the invisible and inaccessible microworld of atoms. In a letter dated December 11, 1867, to his childhood friend Peter Guthrie Tait, professor of natural philosophy (as physics was then called) at the University of Edinburgh, Maxwell wondered what would happen if he could control and redirect molecules at will. Since his fingers were too large and clumsy to grasp them, and his eyes too weak to see them, he invented his little helper, a minuscule and "very observant and neat-fingered being" who could follow and even manipulate individual molecules.

Maxwell designed various special tasks for his gremlin and tried to predict what their macroscopic, observable consequences would be. In particular, he imagined two boxes filled with gas at different temperatures and connected by a trapdoor. The Demon would sit in one of the boxes, next to the door, cowering under a hailstorm of flying molecules, and sort them out—for which reason he is also known as the sorting demon. His job was to open the door only for the occasional exceptionally fast, hotter molecule from the cooler box, allowing it to pass into the hotter box, but shutting the door on all others. The Demon would allow no hot molecules to sneak into the cool box. In this fashion Maxwell caused his diminutive gatekeeper to let heat flow uphill, as it were, from a cold body into a warmer one.

By dint of his prodigious intelligence and dexterity, the goblin could cause things to happen that are never seen to occur in nature, things that seemed able to violate the second law of thermodynamics. Is such a mechanism possible? If not, why not? But if it is, what does it imply about the validity of the second law? Maxwell was not sure of the answers. Deeply troubled by the paradoxical behavior of his creature, he began to discuss it with his friends and colleagues—and his successors have been struggling with it ever since.

The goblin might have remained a half-serious, largely forgotten bit of whimsy had it not been for another Scottish friend of Maxwell's who took a fancy to it and made it immortal. William Thomson, the Glasgow professor of natural philosophy whom Queen Victoria would later elevate to the peerage as Baron Kelvin, appreciated the power of Maxwell's invention more fully than Tait. In 1874 Thomson published an article in which he described what he called "Maxwell's intelligent demon," and the phrase stuck.

By calling the creature a demon, Thomson fixed its image for posterity as a being of endearing levity, sufficiently human to appeal to nonscientists, yet possessed of such fiendish cleverness and agility that it continues to pose troublesome problems for physicists to this day. But precisely by attracting attention through its antics, Maxwell's Demon has also helped to explain how the rules that govern the behavior of heat originate in the unruly motions of atoms.

This sprite is not evil, as Thomson took pains to explain: "The word 'demon,' which originally in Greek meant a supernatural being, has never been properly used to signify a real or ideal personification of malignity." He has less in common with the malevolent cloven-hoofed, pointy-eared spirits of medieval folklore than with the calm, internal voice from which Socrates received reliable advice throughout his life, and which he called his demon. (At his trial, Socrates paid the ultimate tribute to the wisdom of his demon by revealing that it had prevented him from going into politics!) Maxwell's Demon reflects the personality of his creator: He is perfectly good-natured, but his intellectual superiority sometimes renders him profoundly inscrutable.

Maxwell's Demon is unique in world literature. No other fictional creature has a comparable scientific pedigree, no other scientific device such fantastical attributes. The fundamental premise behind his invention—the notion of descending into the world of atoms—was not new. As long ago as 1606, at the very dawn of the scientific revolution, the English polymath Thomas Harriot, Sir Walter Raleigh's science advisor, wrote to the great German astronomer Johannes Kepler, who was inquiring about the cause of the rainbow: "I have now conducted you to the doors of nature's house, where its mysteries lie hidden. If you cannot enter, because the doors are too narrow, then abstract and contract yourself mathematically to an atom, and you will easily enter, and when you have come out again, tell me what miraculous things you saw." Unfortunately this unorthodox advice was ignored by Kepler and the scientists who succeeded him, until Maxwell took it up again.

Who is this Demon of his? Is he merely a toy, a "restless and lovable poltergeist" invented for the sheer fun of it, or is he as serious as over two hundred references in the technical scientific literature would suggest? Is he fantasy or, to the extent that he always obeys the laws governing the motions of atoms, fact? Is he a living being or a machine? Does he belong to our world of complex organisms or the microworld of structureless particles? The cunning Demon manages to live precisely balanced on the line between invention and discovery, between poetry and science. In this age of relativity and uncertainty, such ambiguity is not a flaw—it is, in fact, the hallmark of modernity.

The real significance of the Demon lies neither in his diminutive size nor his superhuman dexterity, but, as Lord Kelvin correctly perceived, in his intelligence. He uses it to process information about his environment, such as the gas around him, in order to accomplish his astonishing feats. It is only now, in the waning years of the twentieth century, that we are beginning to be able to grapple with the physics of information processing and the subtleties of the influence of the environment on inanimate systems. The precocious Demon was a century ahead of his era in the questions he raised.

As it turned out, Maxwell and his colleagues developed thermodynamics to a state of near perfection without the help of the Demon. By 1900 the whole business seemed complete, except for a few loose ends, but then physicists virtually abandoned thermodynamics in pursuit of other interests. They charged off to a two-front war—the attack on the structure of the atom itself, and the assault on the outer reaches of the universe. The tangible world of the here and now was left behind. Thermodynamics was turned over to the chemists and replaced by quantum theory and relativity in the curricula of physics students.

Now, after nearly a century of neglect, thermodynamics is enjoying a dramatic revival. In the search for a universal theory of everything from subatomic particles to the cosmos, Maxwell's enigmatic Demon has assumed an importance far out of proportion with his diminutive stature. Through his paradoxical pranks he has suggested a new interpretation of the second law, and of its relationship to the theory of atomic processes. The characteristics that have led the Demon into the very core of modern physics are his intelligence, his capacity to process information, and above all his habit of blending into his environment instead of standing outside and above it.

In these respects, he mirrors humanity's modern view of itself. We too consider ourselves intelligent, we process oceans of information, and we are finally, painfully beginning to realize that we are willy-nilly part of our own environment. In the words of the venerable American physicist John Wheeler: "Useful as it is under every-day circumstances to say that the world exists 'out there' independent of us, that view can no longer be upheld. There is a strange sense in which this is a *Participatory Universe.*" That is precisely the lesson which the Demon, the first participator on the atomic scale, has tried to teach us all along.

But to understand why he came into being, and how he developed into what he is today, we must go back to his roots in the soil of human experience. The story of the Demon unfolds against the backdrop of the history of thermodynamics—the ancient quest to find meaning in the familiar experiences of warming and cooling, of boiling and freezing, of fire and ice.

And this may be the wily Demon's cleverest trick of all. Even as he leads us forward to the frontier of theoretical physics, he takes us back into history to look for the traces of the evolution of concepts such as heat and energy, without which we could not begin to describe, let alone understand, the complex world in which we live.

WARMTH DISPERSES
AND TIME PASSES

1

INSIDE THE BARREL OF A CANNON: THE NATURE OF HEAT

The laws of thermodynamics smell of their human origin.

—P. W. BRIDGMAN

Warmth, the carrier of comfort and security for human beings, is the primary object of the study of thermodynamics. The vital clue to its true nature was discovered unexpectedly in 1797, when the American Tory spy Benjamin Thompson, Count Rumford, stumbled upon it while manufacturing cannons in Munich. "I was led into these investigations by accident, and in some measure against my will," he wrote to a friend at the time. But in truth, as Louis Pasteur pointed out, chance only favors the prepared mind. Benjamin Thompson's whole career, and indeed even his personality, prepared his mind to appreciate the significance of what he observed.

The immediate purpose of making cannons was not to satisfy scientific curiosity but to kill Frenchmen. As major general and commandant of the police at the court of Carl Theodor, the Duke of Bavaria, Rumford was responsible for the defense of Munich. Two years after the events in question, in 1799, Napoleon was to come to power and bring focus to France's imperial ambitions, but at the time the republican forces of the French Revolution were still fighting the despised Austrians all over the map of Europe. Although Bavaria was nominally neutral in this conflict, Munich, located on the direct route from Paris to Vienna, found itself in the line of fire. As a precaution against future incursions, Rumford decided to upgrade his artillery, and for that purpose ordered the manufacture of heavy brass cannons.

Each barrel, after being cast in the form of a solid cylinder, was brought for boring to the arsenal's machine shop and placed horizontally on the bed of a lathe. A massive screw, continuously kept tight by an attendant, pushed the stationary drill bit—a blade made of hardened steel—into the front end of the barrel with a force of several tons. At the same time, a revolving shaft attached to the back of the barrel turned the entire casting on its own axis at the rate of thirty-two revolutions per minute. The power for turning the cannon was provided by two draft horses harnessed to a winch in a chamber off to the side on a lower level. A system of gears transmitted the motion of the winch up to the lathe's shaft.

The shop was an unlikely laboratory for scientific research. The screech of metal biting into metal combined in an excruciating cacophony with the rumble of the enormous wooden gears and the groaning of the cannon barrel as it rolled on its bed of greased bearings; the stench of steaming draft horses mingled with the acrid smell of freshly cut brass; jagged chips of metal shot from the drill bit into the gloom, which the feeble light from the tiny, soot-covered windows was powerless to dispel.

In this phantasmagorical atmosphere Rumford kept order. He was a slight man, with intense blue eyes and a permanent expression of reticence and mild disapproval. Aloofness was the price for his remarkable power of concentration, which enabled him to reflect upon natural philosophy even in the midst of the clamor of the arsenal shop.

The question on Rumford's mind arose from the naive but compelling curiosity that great scientists have in common with children: What is heat? What is this stuff that is somehow created at the drill bit? Heat and fire and their uses had fascinated him all his life. As a thirteen-year-old, shivering in the drafty house in Woburn outside Boston in which he was born, and which still stands today, he had kept neat, orderly notes on recipes for rockets and firecrackers. Once, when he tried them out, he accidentally set off an explosion and got severely burned. But the mishap didn't deter him: In later life we seldom find him far from cannons, guns, furnaces, fireplaces, lamps, or stoves. Gunpowder fascinated him; heat was his passion.

In his methodical way he undertook to quantify the well-known phenomenon of frictional heat, to find out *how much* heat is produced by metal rubbing against metal. To this end he cast a specially shaped cannon barrel that could be thoroughly insulated against loss of heat, replaced the sharp boring tool with a dull drill bit, and immersed the front part of the gun, where the action was, in a tank full of water. He later reported to the Royal Society in London that after a short while ". . . I perceived, by putting my hand into the water and touching the outside of the cylinder, that Heat was generated; and it was not long before the water which surrounded the cylinder began to be sensibly warm."

Then came the climax: ". . . at 2 hours and 30 minutes it ACTUALLY BOILED. It would be difficult to describe the surprise and astonishment expressed in the countenances of the bystanders, on seeing so large a quantity of cold water heated, and actually made to boil, without any fire." In the candid style of eighteenth-century scientific reports, Rumford continues: "Though there was, in fact, nothing that could justly be considered as surprising in this event, yet I acknowledge fairly that it afforded me a degree of childish pleasure, which, were I ambitious of the reputation of a *grave philosopher,* I ought most certainly rather to hide than to discover." Childish pleasure is something that nourishes and sustains all but the most hidebound scientists, but lack of ambition was not part of Rumford's character.

Even as a youth Benjamin Thompson had stopped at nothing to advance his station in life. At the age of nineteen he married a wealthy widow, fourteen years his senior, from a section of the New Hampshire town of Concord that had formerly been called Rumford. In 1775 he turned spy for the royal governor of the colony, to whom he wrote in a secret ink which he had concocted himself. Once he was arrested by the patriots upon suspicion of treason, but released for lack of evidence. In March 1776, when the British position in Boston became untenable, he abandoned his wife and infant daughter and left for adventures abroad.

For all his opportunism, Rumford was brilliant, hardworking, and extremely orderly. In his writings, from the earliest diaries to his last will, his preoccupation with order is evident. He kept such excellent notes on experiments that he was able to write them up for publication years later, whenever the whirlwind of his professional and social activities allowed a moment of leisure—an unusual and useful skill for a scientist. His personality was perfectly attuned to the emphasis on order, system, and method that characterized the prevailing enlightenment philosophy.

The invention that brought him fame and wealth was an improved fireplace, the so-called "Rumford stove," which produced considerably more heat per pound of wood or coal burned while practically eliminating the annoying and dangerous problem of smoke escaping into the room. To the ordinary householder in eighteenth-century Europe, the man who could prevent this perennial nuisance was both a genius and a savior.

As he watched the water coming to a boil in that abominable machine shop in Munich, Rumford, the tireless inventor, naturally dreamed of putting the impressive spectacle to practical use: ". . . these computations shew further how large a quantity of Heat might be produced . . . merely by the strength of a horse, without either fire, light, combustion, or chemical decomposition; and, in case of necessity, the Heat thus produced might be used in cooking victuals." But in the next sentence Rumford, the scientist, dismissed the idea by pointing out that burning the horse's fodder might provide more heat—and with considerably less trouble. The realization that the power to produce heat somehow traveled from the oats to the horse and thence, via winch, gears, and shaft, to the borer foreshadowed the discovery of the conservation of energy half a century later.

Unable to put frictional heating to immediate use, Rumford nevertheless continued in what he called "abstruse speculations" about its nature. The prevailing theory of heat at the time was that it was an invisible fluid named "caloric" by Antoine Lavoisier, the father of modern chemistry. Phrases like "the flow of heat" and the "heat capacity" of an object remind us to this day that the explanation of heat as an invisible fluid is by no means implausible. Just as water flows downhill, heat moves spontaneously from a hot body into a cooler one, and like buckets of different sizes, solid bodies differ in their ability to store heat. Acceptance of the fluid theory of heat was bolstered by the success of the fluid theory of electricity, of which we still find traces in our words "current," "flow," and "capacity."

In Lavoisier's influential textbook *An Elementary Treatise on Chemistry*, published in 1789, thirty-three basic substances were identified as elements. The list began with caloric and light before continuing with oxygen, nitrogen, and hydrogen. Two decades later the great Swedish chemist Jöns Jakob Berzelius divided matter into ordinary elements and compounds, plus five additional weightless and invisible substances: positive and negative electricity, magnetism, light, and caloric. The chemists, accustomed to dealing with the tangible stuff of the world, endowed the mysterious phenomenon of heat with as much concreteness as they could muster, and called it a fluid.

The fluid theory of heat held that in a hot body the caloric hides in the interstices between atoms. It was supposedly an exquisitely rarified fluid endowed with a self-repelling property analogous to the self-repulsion of the electrical fluid, which, in turn, was deduced from the spectacular display of hair standing on end on the head of an electrically charged child. If the caloric were similarly self-repelling, addition of heat to matter should result in the expansion of the material—which is indeed a universally observed fact: The volume of a fixed quantity of air, for example, expands by a third upon being warmed from the freezing point to the boiling point of water. Frictional heating was explained by the assumption that the process tends to squeeze the materials at the point of immediate contact—the brass of the gun and the steel of the borer, for example. This pressure, the calorists thought, resulted in the expulsion of caloric into the surrounding medium—water, in Rumford's experiment—which caused its temperature to rise.

All this was well known to Rumford, but he was never one to abide blindly by the authority of others; he liked to think things out for himself. Better than anyone, he knew the value of heat, and if he couldn't find a use for the amazing display before him, at least he could try to understand it. Toward the end of his report he writes: "By meditating on the results of all these experiments, we are naturally brought to that great question which has so often been the subject of speculation among philosophers; namely: What is Heat? . . . Is there anything that can with propriety be called *caloric?*"

And then he reminds his readers of the lavishness with which the heat is produced: "We have seen that a very considerable quantity of Heat may be excited in the friction of two metallic surfaces, and given off . . . without interruption or intermission, and without any sign of diminution or exhaustion." As long as the lathe turns, no matter how long that may be, new heat is generated. The quantity of heat in a barrel is evidently inexhaustible, a fact that struck the count with particular force because his attention was perpetually focused on ways of saving and conserving heat.

Finally Rumford comes to a simple conclusion—a short passage that assures his fame long after his duplicity and his stove have been forgotten: "It is hardly necessary to add, that anything which any *insulated* body . . . can continue to furnish *without limitation*, cannot possibly be *a material substance;* and it appears to me to be extremely difficult, if not quite impossible, to form any distinct idea of anything capable of being excited and communicated in the manner the Heat was excited and communicated in these experiments, except it be *Motion*."

If the caloric were a substance, he argues, it would be completely squeezed out of the metal early in the experiment. It is unreasonable to imagine an inexhaustible supply of the stuff stored up in the brass, if for no other reason than that it would have melted the cannon barrel long before. In short, heat cannot be material, and Lavoisier's caloric theory is refuted.

This alternative hypothesis, that heat is some form of motion, had been proposed as far back as the twelfth century. Later, around 1600, Francis Bacon, the evangelist of the scientific method, surmised that "heat itself, its essence and quiddity is Motion and nothing else," and, more specifically, "Heat is a Motion of expansion, not uniformly of the whole body together, but in the smaller parts of it . . . a motion alternative, perpetually quivering, striving, and irritated by percussion . . ." Robert Hooke, a contemporary and rival of Newton's, held a similar opinion: ". . . heat being nothing else but a brisk and vehement agitation of the parts of a body . . ."

In view of such precedents, Rumford cannot claim credit alone for the interpretation of heat as motion. His more modest, but historically crucial, accomplishment was to wrest the study of warmth away from the chemists, who had tried to include heat in their classification of substances, and to restore it to the physicists, who dealt with it in much more abstract terms. Matter and its transmutations are the business of chemistry, motion the concern of physics. Since Newton's laws had by that time been elaborated into a mature theory of motion, physics provided a more hospitable environment for the study of heat than the fledgling science of chemistry could offer. So, inasmuch as "thermo" comes from the Greek for "heat," and "dynamics" is the study of motion, the history of science recognizes Rumford, who put the two together, as one of the principal prophets of thermodynamics.

Heat is motion, Rumford proclaims, and stops. Nowhere else in the annals of science has so portentous a pronouncement ended so abruptly. The enquiring mind fairly clamors for clarification. If heat is motion, what is moving, and where, and how? Can we feel or see or otherwise experience that motion the way we can feel heat? How does it spread? How can we measure it? But Rumford refuses to answer. "I am very far from pretending to know how . . . that particular kind of motion in bodies which has been supposed to constitute heat is excited, continued, and propagated, . . ." he protests, and "I shall not presume to trouble the [reader] with mere conjectures."

By failing to make the connection between motion and atoms, Rumford sealed the fate of his theory, which would be ignored for another half century before being rediscovered and enshrined as fact. In his defense, it must be remembered that the full acceptance of the idea that all matter is composed of atoms, which was anticipated by Maxwell in the nineteenth century, had to await the twentieth. To be sure, the chemist John Dalton would soon introduce units of matter he called atoms, but these were mere bookkeeping devices for tracking proportions of elementary materials in combination, not real, physical particles. Throughout the nineteenth century, the atoms of the chemists and those of the physicists were distinct, independent conceptions. If pressed, Rumford would have been inclined to think of heat in Baconian terms as a kind of wave motion, like that of the crests and troughs of an ocean swell, or of troops of soldiers on parade. While such imagery must have appealed to Rumford's sense of order, it bears no resemblance to the modern understanding of heat. How he would have despised our interpretation of heat as the random, chaotic quivering of molecules!

Rumford's science, like his political outlook, was too totalitarian, his purpose too utilitarian to lead to a true understanding of heat. His approach to nature was more that of a general commanding an army than that of a humble supplicant seeking enlightenment. In a letter to a friend in which he described how he was led to the investigations on frictional heat by accident, he confessed as much:

> I have often thought that I should be perfectly satisfied could I but obtain the exercise of the *authority* [italics in orig.] of a Magician, even though I should not be permitted to know *how* the obedient spirits I should call up performed their business. I can conceive no delight like that of detecting and calling forth into action the hidden powers of nature!—Of binding the Elements in chains, and delivering them over the willing slaves of Man!

Achieving the authority of a magician and delivering chained slaves are not the goals that lead to success in science.

The pursuit of power, not the search for illumination, motivated Rumford. Seven years after his cannon-boring experiment discredited the caloric theory of heat, its inventor Antoine Lavoisier was falsely accused of corruption by the zealots of the Age of Terror and guillotined. Not content with his victory over the unfortunate chemist in the field of science, Rumford settled in Paris and succeeded in boosting both his fame and his fortune one last time by marrying Lavoisier's wealthy widow.

But his triumphs didn't last. After a year the marriage soured, and within a short time the mechanical interpretation of heat was rejected by the scientific community. In the opening decades of the nineteenth century, before the meteoric rise of the science of electricity brought it into the limelight again, physics was eclipsed by progress in chemistry, and Lavoisier's caloric retained its ascendancy for a generation after it should have been abandoned. Acceptance of Rumford's enigmatic proclamation that heat is motion had to await a conceptual revolution as radical as that of Newton himself—the discovery of what we call the first law of thermodynamics—that energy is conserved.

2

THERE'S NO FREE LUNCH: THE ORIGIN OF THE FIRST LAW

[Thermodynamics] is the only physical theory of a general nature of which I am convinced that it will never be overthrown.

—ALBERT EINSTEIN

Centuries before heat was recognized as a kind of motion and a form of energy, inventors dreamed of constructing an engine that would run on its own without need of an external source of fuel, heat, or power—the famed perpetual motion machine. Undeterred by an unbroken record of failure, they persisted in their hopeless search. But although they never found what they were looking for, their efforts led to the discovery of something more fundamental than they had ever imagined. By showing with frustrating certainty that perpetual motion is unattainable, they prepared a stable and enduring foundation for the proposition that would later become the cornerstone of thermodynamics and enshrined as its first law—that energy cannot be created out of nothing. Ironically, their failure to discover perpetual motion turned into the strongest empirical anchor of any theory in the realm of physics.

One of the earliest known descriptions of a perpetual motion machine is buried in the sketchbook of the thirteenth-century master mason Villard de Honnecourt. This miniature cornucopia of medieval thought comprises thirty-three letter-size sheets of yellow parchment between scuffed leather covers, spilling out a jumble of ink drawings in lively confusion: geometrical constructions, sculptural inventions, architectural details, designs of divers machines and floor plans of churches intermingle with exquisite little studies of people and animals. Miraculously, much of the booklet has survived seven and a half centuries of carelessness and pilfering and now lies locked away in a vault of the National Library in Paris.

The first eight pages contain religious material—crucifixes, seated apostles, copies of church statuary. Then comes a surprise. Facing the meticulous drawing of a sculpture of the sumptuously robed personification of the Church, Ecclesia, a whole page is devoted to a hasty sketch of a rickety frame that supports a vertically mounted wooden wheel. The wheel is adorned with seven hammers whose handles are loosely bolted to the outside of the rim in such a way that they pivot freely. A legend in Villard's neat Gothic script explains: "Many a time have master workmen debated how to make a wheel that turns by itself: here is a way to make one with an uneven number of mallets or with quicksilver."

The contraption, another failed design for a perpetual motion machine, is supposed to turn forever because four mallets pulling the wheel down in one direction always overbalance the remaining three on the other side. Whenever a rising hammer reaches the top it is supposed to automatically flop over to the other side, adding its weight to those that are already on the way back down. (The mention of quicksilver refers to an alternative design—often pictured in the later literature on the subject—in which mallets are replaced by cleverly shaped compartments filled with mercury that sloshes back and forth as the wheel turns.) In fact, what happens is that any such wheel, once put in motion, quickly loses its energy to friction, slows down, and stops.

The casual nature of the explanation suggests that Villard was not particularly interested in perpetual motion, nor even claimed to have invented the device. He was a collector and recorder of ideas, not a creative genius like Leonardo da Vinci, who drew much more detailed versions of the same machine two centuries later. The legend beneath the drawing also implies that designs for perpetual motion machines were as common in the thirteenth century as they are today; indeed, precisely the same schemes are rediscovered by generation after generation of eager inventors in a depressing tradition of delusion. And whenever another machine refuses to work, as it must, the fault is ascribed to some minor technical obstacle that is said to be on the verge of being overcome. The quest for perpetual motion is a timeless scientific folly; Leonardo, an experienced engineer, scornfully dismissed it along with the alchemists' search for the philosopher's stone, which turns lead into gold.

The pursuit of perpetual motion is sustained by the absence of a mathematical proof to the contrary. If a scientific proposition is false, a single counterexample proves its falsehood, but if it happens to be true, neither experimental confirmation nor mathematical demonstration can prove that it will always hold: The fact that dropped apples invariably fall does not imply that they will always fall.

But the real motivation of the seekers after perpetual motion lies beyond such logical niceties. It is the same stimulus that in our day inspired the outrageous claims for drawing energy out of solid metals by means of cold fusion. In the hunt for a copious source of energy, the stakes are so immense that they dull skeptical analysis. If Villard's wheel, or cold fusion, could be made to work, poverty would become obsolete and the world would be transformed by a revolution of unimaginable scope. In view of the magnitude of this dream, is it surprising that people would risk frustration, disappointment, and even ridicule in its pursuit?

Belief in perpetual motion is based on an argument similar to the wager with which the philosopher Blaise Pascal justified his belief in God: The odds may be long, but the payoff is infinite. The disproportionate imbalance between cost and reward, Pascal claimed, proves that it would be irrational to decline the wager. The search for perpetual motion thrives on the same reasoning.

As we look at Villard's ungainly engine, Pascal's wager wraps its seductive arms around us and lulls reason to sleep. We dream: Wouldn't it be great if . . . ? What if the physicists have overlooked something, some minuscule loophole, some minor imperfection in the law? No theory has ever been found to hold forever, without modification, so why not . . . ? Has anyone ever really tried it? Clop, clop, clop . . . slowly the wheel begins to turn, creaking ponderously as each mallet in turn swivels on its hinge, falls over, hits the rim, and starts its descent. Through the spokes rotating with gathering speed we glimpse the glittering dream of unlimited energy.

Then rationality reasserts itself. We mutter an incantation we learned in school: The law of conservation of energy guarantees that energy can be neither created nor destroyed. Cold, indifferent science spoils the vision of a golden age. The first law of thermodynamics rudely shakes us awake.

Still dazed, we continue to stare at Villard's sketch. What is this law that stops the wheel? Where does it come from? Who decreed it? In its peremptoriness it resembles the law of universal gravitation that causes apples to fall to the ground, but it is just as mysterious. Three centuries ago, Isaac Newton called the proposition that the earth pulls on the apple at a distance, without the participation of an intervening medium, an absurdity—he knew that his own law defied common sense, and searched in vain for the real cause of gravity. Is the principle of energy conservation any more compelling than the law of gravity?

The law of conservation of energy is not a mathematical truth like the Pythagorean theorem, but the summary of an immeasurable wealth of experience spanning the history of civilization. In a roundabout way it is also a condition necessary for safeguarding the very survival of that civilization. Consider what would happen if by some cunning trick we could outwit nature and get the wheel to turn after all, creating energy out of nothing. By reducing the friction of the bearings, we could free up a bit of excess power and use it to drive a motor. Suppose further that part of the first law of thermodynamics remained unbroken—that energy, once created, may change its form but never disappear.

The energy created by the wheel would accumulate to haunt us like the brooms conjured up by the sorcerer's apprentice. If the wheel were to drive an electric generator, for example, motion would be transformed into electrical energy, and this in turn into heat, or radiant energy, or back into motion, depending on the use we made of the electricity. After every family in the world had erected one of Villard's wheels in its backyard, and used it to heat, cool, and light its home, and to wash its clothes, cook its dinner, and run the TV, and after every factory had harnessed larger versions of the same machine to meet the clamor for the extravagant trappings of modern civilization, none of that newly generated energy would be lost.

Through friction, motion would create heat. Light would be absorbed by matter and turn to heat, like sunlight that strikes bricks and warms them. Sound would travel through the air and peter out—shaking and heating the molecules it encountered. Electricity would meet with resistance, and in the process heat the wires it traversed. Food would be digested and converted into body heat. After countless metamorphoses all energy, unless it is stored, eventually turns into heat and adds its share to the thermal budget of the planet. And thus, before the entire population of the world had satisfied its boundless appetite for energy, the earth would have to cope with a human heat input that would upset its natural energy balance. The result would be disastrously accelerated global warming, thermal expansion of the oceans, polar ice cap melting, a dramatic rise in the ocean level, and incalculable suffering—the infernal antithesis of the longed-for paradise promised by limitless energy.

From this perspective we should be thankful for the failure of Villard's wheel, and regard the first law of thermodynamics in a more favorable light. Where we initially blamed it for rudely spoiling our dream of perpetual motion, we begin to welcome it as the savior from the catastrophe of unlimited energy production—as the guardian of the balance of nature. Considering the myriad ways in which energy regulates our lives, the laws of thermodynamics assume a dominant role that overshadows the significance of human laws.

Fundamental to the thermodynamic laws is the stubborn fact that perpetual motion is impossible. Throughout the turbulent history of the science of heat we meet this realization over and over again—either explicitly or implicitly—as the ultimate rock of experience upon which the entire theoretical edifice is built. For innumerable inventors, the futility of the search for perpetual motion was a source of frustration, but for one little German boy it proved to be the impetus for a momentous discovery.

3

IN SEARCH OF SOUL: CHASING THE FIRST LAW

It is important to realize that in physics today, we have no knowledge of what energy is . . . It is an abstract thing. . . .

—RICHARD FEYNMAN

In view of its role as central organizing principle for science and technology, and therefore, indirectly, for all our industry and commerce, it is curious that the law of conservation of energy—the first law of thermodynamics—is not named after its creator. Newton's laws of mechanics and Maxwell's equations of electromagnetism will keep their discoverers' names alive for all time—but who should get credit for the basic axiom of the science of heat?

The reason for this puzzling usage is that energy conservation represents a classic case of simultaneous discovery. Thomas Kuhn, whose epochal work *The Structure of Scientific Revolutions* inspired a generation of science historians, identified no fewer than twelve men who, between about 1830 and 1850, independently suggested something resembling the modern law of energy conservation. (Count Rumford, however, is not one of them, since he anticipated the first law without stating it. He is therefore a prophet of thermodynamics.)

Chief among the actual discoverers was the German physician Robert Julius Mayer, who was born in 1814, the year of Rumford's death, in the southern German city of Heilbronn. He was ten when he undertook to build a perpetual motion machine. To this end he placed a small waterwheel in the town brook and tried to amplify its power by means of a series of gears, hoping that the tiny wheel, thus fortified, would be able to drive arbitrarily heavy machines. Unlike Villard's device, which had no external inputs, Mayer's was supposed to take a feeble input and multiply it, but the intent was the same: to create energy out of nothing. The failure of the scheme taught him a lesson about the nature of mechanical work that he would never forget.

In university Mayer studied medicine, without, unfortunately, going beyond one semester of physics. Upon passing his state medical exams he decided to indulge his childhood dream of exploring the mysterious East, so in February of 1840, against his father's advice, he signed on as doctor on the merchant ship *Java* out of Rotterdam, bound for the Dutch East Indies. Since his responsibility of caring for the twenty-eight-man crew left plenty of leisure, he applied himself to the study of physiology from the books he had brought along, specializing on the subject of blood.

After the three-month voyage, a chance observation started him on the road to scientific fame:

> . . . a few days after our arrival at the Batavian roads [off Jakarta, capital of Indonesia] there spread in epidemic fashion an acute . . . affection of the lungs. In the copious bloodlettings I performed, the blood let from the vein in the arm had an uncommon redness, so that from the color I could believe I had struck an artery.

And then, following a spectacular mental trajectory, Mayer leaped from this minor medical anomaly to the formulation of a seemingly unrelated physical principle that would eventually be recognized as one of nature's profoundest laws.

We can imagine him in the dim candlelight of his cramped cabin, bent over with the agony of mental labor as perspiration dripped onto the books and papers piled up all around. With the ship at anchor and the crew on shore, no sound, save the gentle smacking of the waves against the hull, disturbed his isolation—he might have been a monk or a prisoner. He was a serious young man whose tiny wire glasses and neatly trimmed chin whiskers made him look older than his twenty-six years. Intense concentration turned down the corners of his wide mouth and knitted his bushy brows. What went through his mind on that muggy voyage a century and a half ago?

To begin with: What prompted Mayer to set off on this intellectual quest in the first place? It so happened that two years earlier the fourth question of his state medical exam in Stuttgart had been "What influence does continued damp and warm weather exert on a person's state of health?" Amazingly, Mayer's answer was preserved and reads, in part, ". . . the blood becomes richer in carbon, darker, and the difference between red and black blood is less." Thus, by a little miracle of historiography, we learn why his observation of bright red blood startled him enough to "capture [his] complete attention": He had expected exactly the opposite from what he found.

Rejecting as incorrect the explanations of blood color he had been taught in medical school, he turned instead to an older, almost forgotten idea. According to Antoine Lavoisier's eighteenth-century theory of animal heat, the body is warmed by oxidation, or combustion, of carbon from food, with the incidental consequence of changing red arterial blood coming from the heart into the darker venous blood that carries the ashes back to the lungs in the form of carbon dioxide. Since less body heat is required in the tropics, Mayer argued, less combustion will take place, less ash will be produced, and the venous blood will retain the bright red color of fresh arterial blood. This explained the observation, yet had he stopped there, Mayer's name would have remained a mere footnote in the history of physiology.

But he did not stop. In fact, he became so obsessed with the problem that he missed out on the sightseeing that had been the original purpose of the trip:

> I . . . inquired little into the distant part of the world, but preferred to stay on board where I could work uninterruptedly, and where for many an hour I felt as it were inspired, such that I can never recall anything similar before or after. A few sudden insights that shot through me . . . were immediately diligently pursued and led in turn to new subjects.

The diligent pursuit of sudden insights distinguishes the genius from the dreamer.

Mayer realized that besides producing heat, the body also performs mechanical work. Recalling his youthful experience with the waterwheel, which taught him that work cannot be created out of nothing, he connected the two functions, and concluded that the body's combustion process is responsible for the production of both heat and work. To complicate matters, he realized that motion, via friction, also produces heat, and that this indirectly generated heat mingles with the heat that comes directly from combustion. Taken together, these thoughts suggested to him that chemical processes in the body produce motion, work, and heat, and that these must all be interconvertible. In this respect, he thought, the body resembles a steam engine, which also burns carbon to produce motion, work, and heat, and so his reflections turned from physiology to physics. He would regret not paying more attention to this subject at the university.

With the idea that motion and heat are interconvertible, Mayer had recovered Rumford's theory, but even before his return from the tropics he knew that a crucial detail was missing. The whole scheme would be worthless without quantification. "A single number has more genuine and permanent value than an expensive library full of hypotheses," as he put it. In particular, he singled out as the irreducible résumé of his thinking the amount of heat that is generated by a fixed amount of work, a number that is today called the *mechanical equivalent of heat*. This awkwardly named quantity is a number that describes the conversion of heat into work, and vice versa: So much work is equivalent to so much heat. It is analogous to the price of eggs: A farmer will trade so many eggs for so many dollars—and back again. The mechanical equivalent of heat measures the price, in work, that we must pay for producing heat. Mayer knew that Rumford had measured the heat produced by drilling, but had had no means of fixing the other side of the equation, the amount of work performed by the horses. The numerical determination of the mechanical equivalent of heat, Mayer felt, was a task of fundamental importance.

What he brought home upon his return to Heilbronn to practice medicine was a confused set of ideas about the convertibility of motion into heat and vice versa, and the determination to quantify these processes somehow, but he had not yet discovered the first law of thermodynamics. Instead of buckling down to the task of actually measuring the mechanical equivalent of heat, which is what an experimentalist might have done, Mayer, the theoretician, continued to hammer away at the subject with logic and occasional flashes of insight. He was struggling painfully toward a monumental insight.

Nearing his goal, Mayer remembered his chemistry. As Lavoisier's theory of animal heat was already on his mind, he recalled the great chemist's principal contribution to science, the law of conservation of matter. Throughout all chemical reactions, the total weight of matter is unchanged—there are new combinations, and dramatic alterations in the appearance of the reagents, but the underlying stuff remains intact. Matter cannot be created or destroyed. Perhaps, thought Mayer, something analogous can be said about heat—but not in the primitive sense of Lavoisier, who had included heat among the elements in the form of an actual, material substance called caloric. What, Mayer wondered, is the nature of that mysterious commodity that begins with a chemical affinity between carbon and oxygen, and then reappears sometimes as motion, sometimes as work, sometimes as heat?

In the end he called the ability to make matter move force. We call it energy, but neither name helps to explain what it is. For Mayer it had a significance far beyond its value as a new scientific concept. What inspired him to devote his entire life, and to some extent even his sanity, to the promulgation of the law of conservation of energy was its religious significance. It was for him nothing less than the answer to the doctrine of materialism he abhorred, and this belief propelled him the rest of the way up his mental trajectory.

The doctrine of materialism denies the existence of a spirit or soul separate from matter. In this view, the world consists of nothing but inert, material objects moving through a vacuum—in 400 B.C. Democritus had called them atoms; today we might call them elementary particles. Materialism and atomism were confounded with atheism by the Roman poet Lucretius, whose lyrical and grandiose exposition of the atomic theory entitled "On the Nature of Things" was a passionate diatribe against superstition and religion. This offended the deeply religious Mayer, who longed to give the soul the same substantial existence afforded to the earth beneath his feet.

The most potent scientific expression of materialism was Lavoisier's law, which granted a form of immortality to matter. But even phenomena that could not obviously be considered material, such as gravitational, electric, and magnetic forces, could be accommodated in the materialistic doctrine: They were described as mere properties of matter, without an independent existence.

Mayer formulated his nascent concept of energy as a weapon against materialism. Here was something beyond matter, something immaterial yet real, an imponderable new essence. Pursuing the analogy with mass, and realizing that energy too is indestructible yet transformable, he welcomed it as a stepping-stone from gross matter to the spiritual world, from body to soul. Later, a quarter century after his voyage on the *Java*, when his worldview had matured and settled, he would enunciate his belief that the world is made of three indestructible components: matter, energy, and the soul. In Mayer's mind, the concept of an immaterial energy, and its conservation, grew out of a firm belief in an immaterial, immortal soul.

In June of 1841, six months after his return home, Mayer sent an account of his ideas to a journal, but the editor ignored it. Its reasoning was more philosophical than scientific, much of its physics was badly wrong, and it lacked the fruits of experiment and observation. In particular, the paper lacked quantification; Mayer didn't even mention the mechanical equivalent of heat. The paper's rejection stung Mayer into action. He immediately began boning up on physics, and nine months later sent off a much improved version of his paper. It was published on his wedding day, May 31, 1842—the high point in the young doctor's life.

Although it still relied more on philosophical speculation than on empirical evidence, and the mathematical physics was still flawed, the second paper reflected the rapid maturation of Mayer's thinking. Mayer even inserted a specific number: his estimate of the mechanical equivalent of heat. The paper's principal conclusion, that "Forces [energies] are thus indestructible, transformable, imponderable objects," came close to the modern statement of the conservation of energy—neglecting only to add explicitly that energy cannot be created any more than it can be destroyed.

The debate over Mayer's place in the history of science has smoldered for a century and a half, but even his harshest critics concede his importance in the determination of the mechanical equivalent of heat. The curmudgeonly Clifford Truesdell wrote, "As the first to attempt any specific use of the idea that heat and work are interconvertible, the tragicomic muse of thermodynamics chose the muzziest of all her muzzy retinue: Robert Mayer."

The reaction of Mayer's contemporaries to his revolutionary paper was silence. Since he lived outside the professional academic world of German physics, and lacked a champion within it to advance his cause, he was unable to make himself heard through normal channels such as conferences, lectures, research collaborations, and letters to colleagues. Nor did the journal he chose for his first paper, the *Annals of Chemistry and Pharmacy*, reach physicists. Some of the members of the august professoriate who had heard of Mayer even went so far as to ridicule the provincial accent and abrupt manner of the emotionally troubled doctor. Most damaging, however, was the fact that while his paper advanced a broad principle uniting physics, chemistry, physiology, and medicine, it made no specific, testable predictions. To be sure, he reported a calculation of the mechanical equivalent of heat, but a number is useless unless it can be compared with another one. Mayer might as well have advanced the speculation that the moon is made of green cheese, and even computed the price of the cheese, for all anyone cared.

A proud, sensitive man, Mayer had counted on early recognition for his discovery; rejection hurt him deeply. Furthermore, in the space of three years one of his sons and two daughters fell ill and died before they reached the age of three. The combined strain proved overwhelming. In 1850, during an attack of insomnia, Mayer jumped out of a third-story window and fell almost thirty feet to the ground. He survived, but the severe injuries he sustained compounded the deterioration of his mental health. He was forced to begin a long series of voluntary and involuntary hospitalizations and even occasional restraint by straitjacket. The scientific world passed him by: In 1863 Poggendorf's authoritative *Dictionary of the History of Science* incorrectly claimed that Mayer had already died—in an insane asylum.

But in the end, the substance of his accomplishment won out over personal obstacles, and professional recognition caught up with him. By the time Mayer died of tuberculosis in 1878, at age sixty-three, he had received honorary degrees and memberships in a number of learned societies, as well as a knighthood in his own country and the Copley medal of the Royal Society of London, the highest scientific honor England had to offer. More important, his jumbled flashes of insight in that sweltering ship's cabin on the other side of the world had crystallized into one of the most fundamental laws of physics.

4

CURRENTS AND WATERFALLS: MEASURING THE MECHANICAL EQUIVALENT OF HEAT

I should like to write "Door meten tot weten" [through measuring to knowing] as a motto above every physics laboratory.

—HEIKE KAMERLINGH ONNES

While the visionary physician Robert Mayer drew powerful, universal conclusions from slender evidence, the pragmatic English brewer James Prescott Joule (pronounced "jool") proceeded in the opposite direction. Without knowledge of Mayer's work, he simultaneously discovered the law of conservation of energy through relentless, painstaking experimentation. For him, scientific research was a matter of measurement, not speculation.

Joule's legendary obsession with precise measurements is celebrated in a story told by William Thomson, Lord Kelvin, the godfather of Maxwell's Demon. They first met at a conference in Oxford where Joule described his ideas on the transformation of mechanical energy into heat. The audience was unimpressed, except for Thomson, who, already a professor at age twenty-three and seven years younger than Joule, alone understood the significance of the older man's work. Thus Joule, unlike Mayer, acquired a powerful advocate in the establishment. A couple of months later, in August 1847, Thomson went off on a grand tour of Europe. One afternoon, while walking along a road near Mont Blanc on the French-Swiss border, he was astonished to meet, coming toward him on the same road, Mr. Joule and his new bride—on their honeymoon. And what was that worthy gentleman doing? Why, he was carrying a long, slender mercury and glass thermometer, related Thomson, and was about to measure the rise in temperature of the water at the bottom of a waterfall, which is predicted by the law of conservation of energy. Not even his own wedding trip could divert the resolute James Joule from his pursuit.

Though the story of the thermometer is probably apocryphal, the accidental meeting did indeed take place, and a waterfall's conversion of motion into heat is real enough, resulting in a measurable warming up of the water. The higher the waterfall, the greater the warming. (The water's crash onto the valley floor converts the orderly downward rush of its molecules into chaotic, random motion—i.e., into heat—without changing the water's total energy.) Fact or fiction, the anecdote illustrates Joule's tireless quest for experimental evidence of energy conservation.

James Joule, who was born on Christmas Eve, 1818, four years after Mayer, was the son and heir of a Manchester brewer whose wealth was sufficient to employ six live-in servants. As a teenager, Joule entered the business and for over twenty years looked after its affairs every day, from nine in the morning until six in the evening. He had been imbued with the scientific spirit by his private tutor, the great John Dalton, founder of the atomic theory of chemistry, and pursued his avocation both at home and at the brewery. Although there is no direct connection between beer and the first law of thermodynamics, the influence of Joule's professional expertise in brewing technology on his scientific work is clearly discernible.

Manchester, which later inspired Friedrich Engels to coin the phrase "Industrial Revolution," was powered by the steam engine, but in the 1830s, when Joule was growing up, scientists were pinning some hopes on the electric motor as a possible alternative. It is not surprising, therefore, that the young man turned his attention in that direction. In the course of trying to develop a more efficient type of motor, Joule noticed that the flow of electricity always produces heat, and decided to quantify the phenomenon. Although this warming is a nuisance in motors and transmission lines, which waste power on the pointless exercise of heating up the atmosphere, it is put to good use in electric water heaters and kitchen ranges. By 1840, at age twenty-two, Joule was able to announce the law that governs the relation between electrical current flow and heating, a formula that today bears his name.

Pursuing the trail backward from the consumption of electricity to its creation, Joule turned from motors to electrical generators. Whether they are powered by steam, water, or even human muscles, generators convert mechanical force into electricity. Since a current can, in turn, be used to produce heat, Joule thus arrived at a way of transforming mechanical effort into heat, and a means of determining the mechanical equivalent of heat.

The mechanical equivalent of heat, as Mayer had also emphasized, is the linchpin of the first law of thermodynamics. It is the fixed, immutable quantity of mechanical effort required to generate a unit of heat—a measure of the price of heat, in units of energy. A discussion of energy conservation without knowledge of this number is a metaphysical exercise akin to a treatise on the economics of brewing without reference to the actual price of a pint of beer. Joule was well aware of the futility of such an exercise.

Having first measured the mechanical equivalent of heat by means of an electrical generator, Joule was determined to generalize his discovery: "I shall lose no time repeating and extending these experiments, being satisfied that the grand agents in nature are, by the Creator's fiat, *indestruc-*

tible; and that whenever mechanical force is expended, an exact equivalent of heat is *always* obtained." The grand agent Joule was beginning to discern vaguely, intuitively, ultimately sustained by his religious convictions, was energy.

One experiment is not enough; one measurement proves nothing unless it is compared with another one. To prove that the mechanical equivalent of heat is a universal number, independent of the details of its measurement, Joule had to find at least one other, independent way of producing heat. He discovered it in the stirring of water, the effect he was supposed to have been pursuing on his honeymoon in the Alps. If heat is motion, and motion heat, agitation should make water warmer—a fact that is common knowledge among sailors, who noticed long ago that ocean water is warmed by the surf. Joule's clever adaptation of this idea joined Rumford's cannon-boring experiment as one of the great classics of the physics of heat.

The mechanical driving force needed for heating was supplied not by horses, but by a weight on a string that descended through a measured distance. The string was wound around the axle of a paddle wheel, and thus set it in motion. The paddle, in turn, was immersed in a pot of cool water, which it stirred up. The weight and the height through which it dropped furnished a precise measure of the mechanical work done, and the amount of water, together with its rise in temperature, yielded the heat produced. The apparatus and its functioning are models of conceptual clarity and simplicity—the scheme is instantly comprehensible. It is all the more surprising that in actual practice the experiment turns out to be excruciatingly difficult. Nature has set the mechanical price of heat so high that even an extraordinary effort yields only a tiny reward of heat. Intuitively we are well aware of the exorbitance of this price: We know that we cannot heat up our tea no matter how vigorously we stir it. How could Joule hope to measure such subtle effects?

In fact, the minuscule rises of temperature in Joule's apparatus were easily swamped by spurious effects such as leaks of heat into the surrounding air, the warming of the thermometer by the experimenter's hand, the loss of energy due to vibrations of the apparatus, and a dozen more. To study such problems, scientists at Cambridge University have recently re-created Joule's experiment, taking exquisite care to assure authenticity in every detail of design, materials, workmanship, procedure, and environment. Early in the course of this work, the historical detectives made the unexpected and troubling discovery that the minuscule heating of the room caused by the physical exertions of the experimenters overwhelmed the expected rise in temperature of the water! And even after they overcame that difficulty, along with many other equally serious ones, they were not able to duplicate Joule's published results.

Blinded by our belief in the superiority of the present over the past, we might conclude that Joule cheated, but then we would miss the point. On the contrary, we must admit that he probably knew things that we don't know, and that cannot be communicated by technical papers. He was the Paganini of the thermometer, and capable of making that instrument perform unimaginable feats. Merely reading it was high art: "And since constant practice enabled me to read off with the naked eye 1/20 of a division, it followed that 1/200 of a degree Fahrenheit was an appreciable temperature." Such precision with the naked eye was and is unheard of. Joule's friend and champion William Thomson called it magical. Besides his uncanny ability to read scales, Joule must have relied tacitly on a whole repertoire of tricks and habits that made him in his day the undisputed leader in the field of thermometry, and that we may never rediscover because they are no longer useful.

The thermometer, Joule's magic wand, was the direct beneficiary of Manchester's love of beer. At that time, the brewing industry in Manchester was undergoing a change from an art into a science, and the chief instrument of that transformation was the thermometer. Fine control over temperature was replacing the traditional knowledge that had been handed down from master brewer to his apprentice to assure success in making beer. Daily exposure to these influences, together with the help of expert instrument makers in his employ, combined to develop in Joule the virtuosity that explains the origin of the legend of his honeymoon. Just as Saint Peter is pictured in medieval paintings and sculptures holding a key as his invariable attribute, Joule cannot be imagined without a thermometer.

The results of his increasingly accurate stirring experiments over a period of several years were in general agreement with his first indirect electrical determination of the mechanical equivalent of heat. Here, at last, was a straightforward experimental confirmation of the law of conservation of energy: If converting a given quantity of work into electrical energy, and then into heat, yields the same result as converting it directly to heat, no energy is lost in the conversion processes, and the law is proven.

By 1850 the law of conservation of energy, together with the numerical formula for expressing heat in mechanical terms, had found a permanent place in the canon of physics. The professional physicists took it over from Mayer and Joule and cast it in rigorous mathematical form. Today the scientific unit for measuring energy, work, and heat is called the joule and is abbreviated *J*. One J corresponds to the energy required to raise a pound of water nine inches off the ground, or to accelerate it to a speed of seven feet per second (walking speed), or to heat it up through a thousandth of a degree Fahrenheit. The minuteness of that increment is a tribute to Joule's unmatched mastery of thermometry.

For all its majestic sweep and newfound numerical precision, the law of conservation of energy—the first law of thermodynamics—is powerless to explain why a cup of tea won't heat up spontaneously. Mayer and Joule discovered the truth about heat, but it was not yet the whole truth. When Mayer sweated in Java, and Joule hiked in the Alps, the seeds of an additional principle had already been languishing for two decades in a remarkable but obscure booklet published by a young army officer in Paris.

5

PARIS IN THE STEAM: THE ROAD TO THE SECOND LAW

Nothing in the whole range of Natural Philosophy is more remarkable than the establishment of general laws by such a process of reasoning.

—WILLIAM THOMSON, LORD KELVIN, ON CARNOT

Like creeks converging to form a river, the independent contributions of several thinkers came together to create the first law of thermodynamics—a powerful, unifying current that flows through all branches of natural science. In contrast, the second law of thermodynamics—which states that heat always flows "downhill," from hotter to cooler—emerged from the deliberate, methodical research of one man: the French military engineer Sadi Carnot. His seminal work predated the first law by two decades, and fell into the middle of that unfortunate epoch in the history of science, roughly the first half of the nineteenth century, when atomism was not yet universally accepted and Lavoisier's appealing but erroneous caloric theory of heat still held sway. This misconception probably cost Carnot the opportunity of joining Galileo, Newton, and Maxwell in the pantheon of physics as the author of the laws of thermodynamics—but his vision was so clear and his purpose so precisely defined that in spite of his shortcomings he deserves the title "founder of thermodynamics."

The French honor their heroes by naming streets after them. When I noticed that the map of Paris lists not one but three places named after Carnot, I decided to seek them out. The most prominent one starts at the Arc de Triomphe, from which twelve roads radiate in all directions. One of these, Avenue Carnot, a wide, tree-lined residential street, slopes downhill to the northwest. In front of No. 2, a sign announces: AV. CARNOT: GENERAL AND STATESMAN 1735–1823. Alas, that was Lazare Carnot, an engineer and cultured man of letters as well as Napoleon's minister of war—the father of the physicist. My next stop was the Porte Dorée, the Golden Gate, on the opposite side of the city near the zoo. Here a somewhat seedy Boulevard Carnot runs alongside the superhighway that defines the boundary of Paris. In a display of redundancy, it also celebrates Carnot the father; soldiers are always valued over scientists. The last entry on my map directed me to a tiny cul-de-sac called Villa Carnot in an unfashionable northern district, but when I got there, it had disappeared—replaced by a glass-and-concrete child-care center. Privately I like to believe that it once honored the man who invented the science of heat.

Sadi Carnot, who was born in 1796 and named after a medieval Persian poet, grew up in comfortable circumstances on the periphery of the court of Napoleon. Early in life, he began to show the signs of a gallant and genteel disposition. Once, when he was four, he watched as Napoleon splashed a rowboat full of timid women by tossing stones at it. Sadi ran up to him, shook his fist, and shouted: "You beastly First Consul, stop teasing the ladies!"

After obtaining a solid education in mathematics, science, languages, and music under the direction of his father, he studied engineering and then joined the army, but at age twenty-four he moved to Paris and retired as a lieutenant on half-pay. He was a stunningly handsome youth, clean-shaven in an age of beards, with gentle, refined features, curved, sensuous lips, a mop of curly hair, and dreamy, almond-shaped eyes full of intelligence and warmth. His younger brother Hippolyte reported that he loved dancing, literature, and the arts, and that he was an excellent violinist. He never married and valued his privacy—despising, for example, the attention he received on account of his famous father. The work to which he devoted the full energy of his tragically short life was an analysis of the steam engine.

In an intensive study of industrial development, which he pursued not only in academic libraries but also on the floors of factories and workshops, Carnot had become aware of the significance of the steam engine. He knew that its perfection over the course of fifty years had come almost exclusively at the hands of English and Scottish engineers beginning with the famous James Watt; French contributions had been virtually nonexistent. The steam engine had become so dominant an influence on England's industry that removing it might well, in Carnot's words, ". . . annihilate that colossal power." Englishmen, who like the French value their military heroes over engineers, and military prowess over industrial capacity, would have winced at his guess that "the destruction of her navy, which she considers her strongest defense, would perhaps be less fatal [than the loss of the steam engine]." Carnot set out to make up for the failure of his countrymen to participate in the development of steam power.

In 1823 he was ready to publish what he had discovered. Before putting pen to paper, he had adopted two guidelines, each admirable in its own right, but fatal in combination: By neatly canceling each other out, they condemned his book to almost total oblivion. First, he decided to address himself to the general public rather than an audience of scientists and engineers. This decision establishes the book, which much later assumed its rightful place among the classics of science, as the last member of a noble tradition. Galileo himself had started the trend by writing in popular Italian instead of Latin, by keeping mathematical details to a minimum, and by perfecting a lively literary style. Galileo's writings were enormously influential, but after the time of Newton another genre, densely mathematical in content and highly professional in tone, had become predominant, particularly in physics.

Carnot's second guideline, and the essence of his greatness, was to embrace generality. Inspired by his father, who had written a successful book on the analysis of simple mechanical machines, Carnot undertook to develop a general theory of steam engines that would rise above the practical questions of design and materials that were of immediate interest to engineers.

A popular explanation of the advantages of the steam engine, or a general treatise on the theory of extracting work from heat, might have made its mark. But the public was too unsophisticated to understand a general theory, and technical people too contemptuous to bother with what seemed to be a popularization of a complex subject. By trying to address two audiences at once, Carnot excluded both. His *Reflections on the Motive Power of Fire* received only one, albeit enthusiastic, review, and a decade later, three years after its author's death at thirty-six, one single citation in a science text alone bore the burden of keeping his memory alive.

Carnot's book begins with the broadside "Everyone knows that heat can produce motion," and launches into an ode to heat:

> It causes the agitations of the atmosphere, the ascension of clouds, the fall of rains and of meteors, the currents of water which channel the surface of the globe, and of which man has thus far only employed but a small portion. Even earthquakes and volcanic eruptions are the result of heat. From this immense reservoir we may draw the moving force necessary for our purposes.

The book proceeds with a paean to the steam engine as an instrument for industry and transportation. As soon as the reader is drawn in, Carnot's real purpose emerges. He laments the fact that hitherto all improvements of the steam engine have come about in a haphazard way, as a result of tinkering with the design, the materials used in construction, and the types of fuels. What is needed instead, he writes, is a general theory.

On page six of the *Reflections* we come upon the paragraph that marks the birth of thermodynamics. Although Carnot did not live to attain his noble goal, he knew exactly what he was striving for, and his words ring as true today as they did when he set them down over a hundred and seventy years ago:

> In order to consider in the most general way the principle of the production of motion by heat, it must be considered independently of any mechanism or any particular agent. It is necessary to establish principles applicable not only to steam engines but to all imaginable heat-engines, whatever the working substance and whatever the method by which it is operated.

Carnot's intention was as precise as Galileo's in initiating the study of motion, Newton's in founding the science of mechanics, and Maxwell's in creating the theory of electromagnetism. If his name is less well known than theirs, it is because he did not succeed as they did, but in view of the subsequent importance of thermodynamics, he surely deserves to have a Parisian street named after him, or at least an alley.

Carnot set out to develop an abstract theory of how heat can be converted into work (by a steam engine, for example) and work into heat (by Rumford's boring machine) without taking into account the details of the mechanisms. Where Galileo had invented the concept of acceleration to apply to planets, cannonballs, and motes of dust, and Newton had established the laws of motion for stars, apples, and atoms, all without reference to shapes, sizes, or compositions, Carnot wanted to find the rules for the operation of teakettles, steam engines, and volcanoes without reference to their designs.

The two unquestioned axioms upon which he based his thinking were the commonly acknowledged impossibility of perpetual motion, and Lavoisier's erroneous caloric theory of heat. Less than thirty years had elapsed since Rumford's cannon-boring experiment, and the chemist's mistaken theory was still in vogue. Carnot's genius lay in achieving as much as he did with the flawed tools at his command.

His principal insights illustrate with exemplary clarity the difference between the approaches of the scientist and the engineer. In the analysis of the steam engine, the quantity of interest is *efficiency,* measured, say, by the output of work per ton of fuel consumed. In an effort to improve efficiency, pragmatic engineers look for ways of saving fuel, and of maximizing the useful work produced. They systematically attempt to eliminate heat loss by improving design and materials, avoiding useless vibrations, and experimenting with different sizes of machines to find the best compromise between efficiency and unwieldiness. Decades of this kind of tinkering had succeeded in increasing the efficiency of steam engines, but by Carnot's time it had not reached beyond a level of about 6 percent, with 94 percent of the burned fuel being wasted.

Carnot thought that a more fundamental approach to efficiency would be more successful. Sitting in the quiet of his study, far from the noise and stench of the factory, and removed from the pressure to increase productivity and improve profits, he thought the process through in abstract terms. Since heat and its motive power supplied the driving force for the engine's whole system, he tried, in his imagination, to follow the flow of the caloric as it emerged from the coal fire and swirled around the parts of the machine. In this way he came to a startling realization that had eluded everyone before him. He noticed that every heat engine, regardless of design or quality, discards heat into the surrounding environment. In the case of the steam engine, it escapes through the chimney into the air, and by way of cooling water into nearby streams. (In our day, automobiles eject heat through the radiator and the exhaust pipe, atomic reactors send it up their cooling towers, and fossil fuel plants pour vast amounts of it out their smokestacks.) No matter how hard the engineers tried, they couldn't seem to eliminate what they regarded as the steam engine's massive waste of heat.

Making a virtue of necessity, Carnot concluded that the loss of heat to the environment was not just a nuisance, but a natural by-product of the harnessing of the motive power of heat. He compared the caloric to water, and the steam engine to a water mill. The fall of the caloric from a place of high temperature—the boiler—to one of lower temperature—the environment—produced work. In this way he came to regard the rejection of heat as an inevitable phenomenon, as natural as the outflow below a water mill. After going through innumerable examples in his mind, he arrived, via the process of induction, at a new general law: It is impossible to extract work from heat without at the same time discarding some heat.

Carnot himself did not appreciate the importance of this new law of nature, but in time it would be recognized as rivaling the first law of thermodynamics—the law of conservation of energy—in significance. In fact, improbable as it may seem, Carnot's insight would be found to be equivalent to the much simpler notion that heat flows downhill, and canonized as the second law of thermodynamics. Thus it happened that in an indirect, roundabout way, the second law was discovered before the first.

Carnot opened the eyes of engineers to a new way of seeing the steam engine. It is not simply, as had always been believed, a machine for converting heat into work, but a mechanical device interposed between two thermal reservoirs—the first, at a high temperature, to supply heat, and the second, at a lower temperature, to receive the discarded heat. It has not one but two outputs: one of work, the other of heat. Designers of steam engines would henceforth have to pay as much attention to the cooling mechanisms as they did to the furnaces.

Carnot's fresh understanding of the basic scheme of operation of steam engines, which demonstrated the power of his detached, abstract point of view, was followed by a second, even more radical insight. Carnot wondered what would happen if the machine ran backward. Instead of producing work, it would consume work, all the while transporting heat from the cooler reservoir to the hot one. It would, in fact, become a refrigerator. (In the familiar household version, power is fed in through the electric plug, and heat, taken out of the interior, is released into the kitchen through the dust-clogged grill in the back.) If the steam engine is like a turbine driven by water coursing down a slope, a refrigerator works like a pump for returning the water back to the top of the mountain.

Since Carnot was not only a scientist, but also a practical engineer, he was fully aware of the fact that both steam engines and refrigerators entail real losses of power due to unwanted radiation, noise, friction, and a host of other nuisances. But, he argued, let us imagine that these have been completely eliminated by inspired tinkering. What is left, then, is an ideal, theoretical machine that he called *reversible.* By an infinitesimal intervention—a flick of the wrist, as it were—it can be turned from engine to refrigerator and back again, and it works perfectly in either direction. Its imaginary hydroelectric analog is a water turbine driven by water from a mountain lake, which charges a battery, reverses at the push of a button, and, powered by the battery, pumps the water back to the lake—all without sustaining any losses of energy.

The novelty of Carnot's hypothetical reversible engine, which has played a central role in all subsequent thermodynamic thinking, was that it differentiates sharply between unnecessary losses of heat, which are annoying but fixable, and the necessary release of heat that accompanies the production of work even in perfect, reversible machines. Until Carnot made this crucial distinction, engineers invariably lumped the two together and attacked both as enemies of efficiency. Carnot proved that most of their effort was doomed to failure.

The words "reversible" and "irreversible" hint at the wonderful connection that would later be established between thermodynamics and the concept of time. Carnot had no way of foreseeing how his hard-nosed engineering considerations could end up as fundamental concepts in the arsenals of philosophers.

Carnot's third great contribution—after the realization that heat *must* be rejected and the invention of the reversible engine—was by far the most surprising. By careful logic, he managed to prove what has become known as Carnot's theorem, which fortunately survived the demise of the caloric theory. The theorem states that the efficiency of a reversible engine depends only on the temperatures of the two heat reservoirs between which it operates. The magnitude of this discovery can hardly be overstated.

In practical terms this means that once the temperature of the hot reservoir (the fire in a coal plant, or the core of a nuclear reactor, or the cylinder in an automobile engine after the fuel mixture has ignited) is fixed, and the temperature of the cold reservoir (of the surrounding air or a nearby body of water) has also been determined, then the optimum efficiency of the engine is immutably determined and *cannot be increased.*

A different way of stating Carnot's surprising theorem is that all reversible engines operating between the same two temperatures have the same efficiency. Their mechanical designs, the materials used in their manufacture and operation, their size and shape, whether they deliver power mechanically or electrically or chemically—none of these details matters in the least. Carnot found a ceiling above which engineers simply could not improve the efficiency of their machines. Striving for reversibility was the best they could hope for; beyond that the laws of nature forbade further improvement. Pure physics won a stunning victory over engineering practice.

Carnot's theorem implies that the engine with the greatest temperature difference is the best. Just as a hydroelectric generator works best when the mountain is high and the valley deep, a heat engine is most efficient when the inside is hot and the outside, where waste heat is dumped, cold. Your car, for example, is in principle a tad more efficient in the winter than in the summer—provided the temperature of the burning gasoline is the same. But more significant than this minute variation is the depressing fact, as unavoidable as gravity, that about two thirds of the fuel is wasted.

Carnot's theorem is the most impressive example in all of physics of the triumph of theory over praxis, of abstraction over experience. It stands as a monument to the value of pure science—not just to the intellectual endeavor of understanding nature, but directly to the practice of engineering. Thus it establishes science as the very heart of industry and commerce.

Unhappily, Sadi Carnot never got to reap the fruits of his pioneering work. In part this was due to the peculiar way in which he advanced his ideas, but there is probably another more poignant reason as well. Just as his booklet appeared in print, Carnot was beginning to have doubts about its underlying assumptions. Among his papers there are notes proving that he was close to rejecting the caloric theory, and that he even estimated a rough value of the mechanical equivalent of heat twenty years before Mayer and Joule. It seems, in short, that after enunciating a rudimentary form of the Second Law of Thermodynamics, he was about to discover the first.

But in his mind the two were contradictory. There seemed to be two quite different ways to analyze a steam engine. If energy is conserved, as the first law decrees, then the heat produced by the fire is converted into the work performed by the engine. If, on the other hand, the production of work were due merely to the fall of caloric from one reservoir to another, as Carnot erroneously believed, no heat would be actually used up. Which one is true, *conversion* of heat, or *conservation* of heat? If he had lived just a little longer than he did, he would have learned, possibly by himself, that this quandary, which he perceived as a fatal flaw in his understanding of heat, can be resolved with the introduction of the concept of energy, of which heat is merely one form.

After the *Reflections* were published, Carnot did nothing to promote its success, and stopped publishing. For a year he was forced to return to active duty, but then he retired permanently and devoted himself to research. In June of 1832, he was in Paris during an anti-government riot. When he came upon a drunken officer galloping down the street assaulting people with his saber, he reacted as courageously as he had as a toddler in Napoleon's garden: He knocked the ruffian off his horse and threw him into the gutter. Two weeks later Carnot caught scarlet fever, which spread to his brain, and he was taken to the country, where his brother Hippolyte helped to nurse him back to health.

But in August, while he was still frail, a cholera epidemic ravaged the city, caught Sadi, and killed him in a matter of hours. He was only thirty-six years old.

As was the custom with cholera victims, his clothes, his personal effects, and almost all his papers were burned. A bundle of notes was all that survived to document Carnot's intellectual progress after his book. (But documents can outlast their authors: In 1966 a manuscript by Carnot, preceding the *Reflections*, was found and published with a delay of almost a century and a half.)

Six years after Carnot's premature death, Mayer wrote his great theoretical paper on energy *conservation;* a year later Joule produced his unheralded experimental proof of energy *conversion*. Finally, at the mid-century mark, the seemingly contradictory discoveries of the three men were reconciled with astonishing ease.

6

IT'S ALL DOWNHILL: THE SECOND LAW OF THERMODYNAMICS

Once or twice I have been provoked and asked the company how many of them could describe the Second Law of Thermodynamics. The response was cold: It was also negative. Yet I was asking something which is about the scientific equivalent of "Have you read a work of Shakespeare's?"

—C. P. SNOW

After the reunification of the country in 1990, the German Physical Society decided to move its headquarters to a historic mansion in the former East Berlin. A handsome bronze plaque explains why: "In this House," it reads, "the first Institute of Physics in Germany was founded and directed by GUSTAV MAGNUS from 1842 until 1870. This plaque was installed in his memory and that of his colleagues and students A. von Baeyer, E. du Boys-Reymond, R. Clausius, P. von Groth, H. von Helmholtz, G. Kirchhoff, A. Kundt, E. Sarasin, J. Tyndall, and E. Warburg by the German Physical Society in 1930." As I write this book, I am pleased by this association of my great-grandfather Adolf von Baeyer, who went on to win a Nobel Prize in chemistry for the synthesis of indigo, with Rudolf Clausius.

The man in whose house they met, Gustav Magnus, was a professor of physics, and an inspiring teacher. He is known principally for the discovery of the effect that bears his name and that causes spinning baseballs to curve, but the establishment of his school of science was a far greater contribution—not just to the profession, but to society at large. The rest of the plaque reads like the roster of a scientific Dream Team.

Thermodynamics figured prominently in the careers of most of the Institute's members. Magnus himself was instrumental in establishing the mechanism by which gases conduct heat. John Tyndall became the professor of natural philosophy at the Royal Institution in London, which had been founded by Count Rumford. Perhaps for that reason, he was a stalwart defender of that infamous adventurer. By reputation Tyndall was the most accomplished popularizer of physics of his generation. He met Robert Mayer and exerted his considerable influence to promote recognition of the good doctor's work, especially through his 1863 book *Heat Considered as a Mode of Motion*, which went through numerous editions and became a classic. Emil Warburg and August Kundt, friends and colleagues at the University of Strasbourg, corroborated the predictions about heat transport that James Clerk Maxwell had made on the basis of the atomic theory.

Hermann von Helmholtz, who would later be called, in jest and admiration, the Reich Chancellor of German Physics, was only twenty-six years old and had just become aware of Joule's work when on July 23, 1847, he gave a famous lecture to the Berlin Physical Society on the conservation of energy. He coined the phrase "Principle of Conservation of Energy" and proceeded to do what Mayer couldn't quite achieve, and Joule never tried—construct a full mathematical formulation of the first law, as applied to mechanics, heat, electricity, magnetism, physical chemistry, and astronomy. He used it to clear up old puzzles and propose new, testable mathematical relations. So compelling was Helmholtz's paper that he is often cited as the discoverer of the First Law of Thermodynamics, an attribution that is manifestly unfair to Mayer and Joule, but correctly reflects the fact that he was responsible for presenting the law in the austere language of theoretical physics.

The most prominent thermodynamicist listed on the plaque in Berlin was Rudolf Clausius, who removed the contradictions inherent in Carnot's work and cast the corrected results into crisp, mathematical terms acceptable to the community of physicists. He, too, is sometimes called the founder of thermodynamics. Whereas Carnot attempted to discover the general principles of thermodynamics by induction from his detailed understanding of steam engines, Clausius proceeded in the opposite direction. Adopting Carnot's propositions as axioms, he modified and strengthened them by means of specific, experimentally verifiable deductions. In Carnot's synthesis and Clausius's analysis, the second law of thermodynamics found its twin pillars of support.

Rudolf Clausius was a typical German professor in the scholarly tradition of the nineteenth century. His famous paper on the theory of heat in 1850 led to successive university positions in Berlin, Zurich, Würzburg, and finally Bonn, where he served in the university administration for many years. Although the Englishman John Tyndall was a good friend and teammate on the Dream Team, and although, in 1879, Clausius won the Royal Society's coveted Copley medal, eight years after Mayer, he was a chauvinistic defender of Germany's scientific reputation. Several times he became involved in priority battles, most notably in the bitter feud that erupted in the late 1860s over the competing claims of Mayer and Joule for precedence in determining the mechanical equivalent of heat. Clausius pushed the argument beyond the issue of fairness to the discoverers, and elevated it to a question of German national pride.

But all that lay in the future when Clausius tackled the theory of heat in 1850. He had the advantage of hindsight over Carnot: The caloric theory had been soundly refuted by Mayer and Joule, and the first law enthroned in its exalted position by Helmholtz. But Clausius realized that something was still missing from the theory of heat. Why, for example, does a cup of tea always cool down instead of heating up?

Clausius had heard of the little book in which Carnot had initiated the search for a complete, consistent theory of heat, but had been unable to find it. Instead, he had studied its results indirectly in an article by the French engineer Emile Clapeyron—the sole scientific reference to the *Reflections.* In this way Clausius rediscovered the contradiction that had stopped Carnot in his tracks. Since steam engines must discard some heat, it seems that they simply employ it as a working fluid without using it up. Heat, in other words, seems to be *conserved.* On the other hand, the first law states that heat is used up, or *converted* in the production of work. Clausius saw the dilemma—and promptly resolved it.

The compromise he proposed was that a steam engine should be regarded as a device that absorbs heat from a hot reservoir and converts some of it into work (as Joule claimed), and at the same time conserves the rest by moving it to the cold reservoir (as Carnot believed). Carnot had not been able to find this simple way out of his paradox as long as he had held to the mistaken notion that heat, in the form of caloric, is an indestructible substance. The low efficiency of steam engines in his time converted such a small fraction of the energy into work—no more than a few percent—that the torrent of heat cascading through the machine seemed undiminished to Carnot. Had steam engines been dramatically more efficient, he would surely have noticed the disappearance of heat from the energy budget.

From his new perspective, Clausius tackled Carnot's fundamental proposition that the dumping of heat into a cold reservoir is both necessary and inevitable. Carnot had derived it from the indestructibility of the caloric, a notion that was no longer tenable. Clausius believed in the truth of Carnot's proposition, but rejected its proof. What was he to do with it?

In science, a truth that cannot be proved is called a law. When Isaac Newton despaired of deriving the equation governing gravitation from more plausible premises ("I have not yet disclosed the cause of gravity . . . since I could not understand it") he installed it as a law that reigned supreme for a quarter of a millennium: "the universal law of gravitation," or "the law of universal gravitation," which schoolchildren learn by heart. (It remained unexplained until Einstein derived it from his general relativity theory in 1915.)

In the same spirit, Clausius accepted Carnot's claim that all heat engines must discard heat, and called it the second law of thermodynamics. Only experience would show whether it was true or false.

Laws of nature cannot be proved or derived, but they can be recast into different forms. Carnot's version is immediately applicable to engineering practice, but hopelessly remote from everyday experience. Who, besides mechanical engineers, worries about the cooling provisions of a heat engine? To bring it closer to ordinary experience, Clausius managed to find a much more intuitively appealing but logically equivalent formulation of the second law.

Assume, he argued, using the classic gambit of the counterpositive proof, that contrary to Carnot's axiom, there *does* exist an engine that wastes no heat—a device, say, that somehow uses the heat discarded by a refrigerator and converts it to electricity. If that current is in turn used to run the refrigerator, the two machines, considered as a unit, would transport heat from the freezer into the kitchen *without an external power source.* This "perpetual refrigerator" would revolutionize the food industry just as Villard's wheel would have changed the power industry. But of course such a wonderful device does not exist: Heat flows downhill without external help, never uphill. Therefore, the premise must be false: There is no engine without waste, and Carnot's axiom is true.

By this kind of argument, Clausius showed that the second law of thermodynamics is equivalent to the simple statement *Heat flows naturally from hot to cold, but not the other way around.* (By "naturally" he meant by itself, or without the expenditure of work.) In this form, the law is obvious to us all. When you buy a cup of coffee and a container of juice on the way to your office, have you ever taken them out of the paper sack to find that the coffee has stolen heat from the juice, making the coffee even more pipingly hot and the juice more refreshingly chilly? Never. Sadly, you always end up with colder coffee and warmer juice.

Clausius's astonishing feat of reasoning amounts to this: If you believe that heat only flows downhill (and who doesn't?) then you are forced by inexorable logic to accept the disappointing conclusions that you cannot recycle the heat discarded by your car's radiator to improve your gas mileage, and that your electric utility cannot save money by recycling the heat from its cooling towers.

That the conclusion is as difficult to accept as the premise is obvious only serves to accentuate the ingenuity of the argument. Why, we ask, cannot the hot exhaust from a heat engine be put to a useful purpose, contrary to Carnot's claim? Here, as is often the case in science, qualitative reasoning is not enough—one is forced to become quantitative. It turns out to be quite true that a few houses could be heated with the hot air that escapes from the cooling towers of a large oil-fired electric plant, and in some cases this is actually done. It is also true that a small lamp might be powered with the heat derived from a car's radiator, without affecting the engine's efficiency. But that's about as far as you can go. If a massive converter were installed in the cooling circuit of a heat engine for the purpose of capturing a significant fraction of the wasted heat, it would necessarily block the flow of heat, raise the temperature of the coolant, impede the smooth functioning of the engine, and lower its efficiency. The net result would be a loss, not a gain, in useful energy extracted from the fuel.

The true significance of Carnot's form of the second law can be appreciated by turning away from complicated realistic examples such as power plants and automobiles and considering instead the simple, hypothetical, and extreme case of an engine that is constructed so that it recovers all the heat it discards. Suppose we built an electric generator locked up in a closed box and submerged deep in a huge, warm, tropical lake, with electric wires connecting it to the shore. Could this contraption, which needs no access to a cold reservoir because it uses up 100 percent of what it takes in, convert a portion of the lake's immense store of molecular motion into electricity? Short reflection reveals that in principle the actual temperature of the lake doesn't make any difference, because even a normal, cool lake contains a vast quantity of thermal energy. Is it possible, in other words, to draw power out of the ocean or the air if there is no cool reservoir available for accepting waste heat—no place for heat to "flow" to?

The answer, according to the second law, is no, and in this extreme case it is easier to accept. You can't build a machine for extracting energy out of the ocean or the air any more than you can build a perpetual motion machine! The proof of this disappointing conclusion comes less from failed attempts to beat the second law than from countless successful applications of the theory of thermodynamics. By its means we can understand how refrigerators keep food cool, how the earth's hot interior powers the rearrangements of its crust, how the sun provides all living things with warmth, how individual cells convert food into energy, how photosynthesis works, how the universe expands, how solar cells and nuclear reactors produce electricity, how teacups cool. In every case, the second law furnishes not only qualitative explanations but quantitative, testable predictions. The totality of that experience convinces us, beyond the possibility of a doubt, that the second law holds.

Most of our experience with thermodynamics is derived from our own world with its macroscopic scale of sizes and weights. But what happens in the atomic realm, which until recently was not even accessible to our senses? The first law—energy conservation—has been shown to hold true there, and even down to the region of quarks at the most fundamental level of matter. But what about the second law? Does it even apply to atoms and molecules? To answer that question, Maxwell would invent his Demon, but not until 1867, seventeen years after Clausius first began putting the science of heat on a systematic footing.

In spite of the far-reaching implications of the second law, Clausius's reformulation came close to trivializing it. He admitted, in effect, that after all the brilliant insights of Rumford, Carnot, Mayer, Joule, and Kelvin, and others too numerous to mention, science could do no better than to decree that a cup of tea cools. Heat flows out of the cup and into the air, not the other way around, because the second law says so—and in 1850 no one knew where that law came from. It was as obvious to the founders of thermodynamics that the deep roots of the second law were yet to be discovered as it was to Newton that the law of gravity required further explanation.

Before those roots could be exposed, one more obstacle had to be overcome: The second law had to be quantified. When heat is drawn from the boiler of a steam engine, some of it is converted into work and the rest is rejected—but how much is *some?* While Carnot's version of the second law decreed that waste is inevitable—that engines cannot help being inefficient—it failed to describe this prediction in numerical terms.

Immediately after presenting the second law in clear, simple words, Clausius set to work to recast it once more—into the mathematical language that is appropriate for making calculations. He succeeded, but he found that there was a price to be paid. Hard as he tried, he couldn't couch his discovery entirely in terms of energy, but was forced to introduce a new concept into the description of heat—an abstract, theoretical quantity for which he coined the word "entropy."

7

BRIEFER IS BETTER: THE INVENTION OF ENTROPY

The law that entropy always increases—the second law of thermodynamics—holds, I think, the supreme position among the laws of Nature.

—SIR ARTHUR EDDINGTON

The conventional perception of the scientific method as a dialogue between empirical observation and mathematical analysis, between facts and theories, seems flat and two-dimensional to the extent that it leaves out the larger intellectual landscape in which the scientific enterprise is embedded. The physicist Gerald Holton has enriched the vocabulary of the history of science, and our understanding of the scientific method, by introducing a third dimension, which he calls the *thematic content of science.* In his view, a few simple themes—unspoken assumptions and intuitively held prejudices that originate outside science—underlie all scientific thought. Although these presuppositions may eventually be modified or discarded on the basis of the evidence, they are initially dictated neither by the facts nor by the theory. In Holton's perspective, science is revealed as a tangle of experimental data, logical analysis, and unfounded preconceptions.

Among the persistent themes of physics, Holton included the search for invariants—quantities that remain unchanged throughout the unfolding of some physical process. Robert Mayer, for example, believed in the conservation of energy in large part because he saw it as analogous to the immortality of the soul. For James Joule the idea of energy conservation was inspired by a belief in the divine order of the universe. Neither man's philosophical and religious prejudices had compelling scientific bases, but nevertheless provided motivation and direction for research.

A different theme of great power in the past, as well as in today's search for the so-called final theory of physics, is aptly called "parsimony" by Holton. The principle of parsimony values a theory's ability to compress a maximum of information into a minimum of formalism. Einstein's celebrated $E = mc^2$ derives part of its well-deserved fame from the astonishing wealth of meaning it packs into its tiny frame. Maxwell's equations, the rules of quantum mechanics, and even the basic equations of the general theory of relativity similarly satisfy the parsimony requirement of a fundamental theory: They are compact enough to fit on a T-shirt. By way of contrast, the human genome project, requiring the quantification of hundreds of thousands of genetic sequences, represents the very antithesis of parsimony.

Einstein implicitly included parsimony in "the noblest aim of all theory," which, he thought, was "to make [the] irreducible elements as simple and as few in number as is possible, without having to renounce the adequate representation of any empirical content." Theories, in other words, should be made as simple as possible, but not simpler.

Rudolf Clausius was guided by the principle of parsimony as he searched throughout his career for better ways to express the laws of thermodynamics. Like a poet pursuing the perfect phrase he returned to the subject over and over again until he succeeded: His final formulation is a monument to parsimony.

Accepting Joule's determination of the mechanical equivalent of heat, and correcting Carnot's results, Clausius announced in 1850 that (1) Energy is conserved, and (2) Heat flows naturally from hot to cold. These statements of the first two laws of thermodynamics were admittedly succinct, but they offended Clausius's German sense of order by their lack of parallelism. The first proposition had been turned into a quantitative, mathematical equation (notably by Helmholtz), but the second remained an obvious and rather commonplace observation without mathematical content. To remove this asymmetry, Clausius published a second paper in 1854 under the title "On a different form of the second law of thermodynamics."

In his reformulation, Clausius set aside his own simple version of the second law (heat flows downhill) and returned to Carnot's older statement that heat engines must waste heat. For simplicity, Clausius focused on perfect, reversible machines, which do not suffer from friction. He asked the quantitative question: If a reversible engine, operating between two fixed temperatures, takes in a given amount of heat, how much will be converted into work and how much ejected at the lower temperature? The law of energy conservation demands that the sum of those two unknown quantities should equal the energy used up, and the second law that neither of them is zero—but to fix their individual values, a new equation was needed.

When Clausius computed the properties of reversible engines he found a simple pattern: The ratio between the large, given amount of input heat and the high temperature of the combustion chamber always equaled the corresponding ratio of the wasted heat to the low temperature of the coolant. This striking and symmetrical relationship worked only with the new "absolute" temperature scale, based on physical principles rather than arbitrary conventions, which Joule's friend William Thomson, later Lord

Kelvin, had just introduced. When the temperature was expressed in degrees Fahrenheit or centigrade, it failed. Furthermore, the formula also broke down for irreversible engines; it seemed to be a hallmark of reversibility.

Clausius knew that the discovery of some simple mathematical relationship is often the first hint of a momentous insight. Archimedes created the pattern for theoretical physics when he stated the law of the lever: For a beam, such as a balanced seesaw, the product of weight times distance to the fulcrum on one side equals the corresponding product on the other side. Two millennia later Newton based his theory of motion on the discovery that for two interacting bodies, the product of mass times acceleration is always the same for both bodies. In the same spirit, the constancy of the ratio of heat to temperature turned out to be the key to the quantification of thermodynamics.

The individual elements of the equations of physics at first bear no relationship to each other. Weight and lever arm, for example, are independent properties, measured by different devices—a scale and a yardstick—in different units—pounds and inches. Similarly the mass and acceleration of an apple are separate and autonomous quantities. And yet, the law of the lever, and Newton's law of motion, manage to arrange these seemingly unrelated attributes into simple but powerful combinations.

On the face of it, the ratio "heat over temperature" seems as far-fetched as the products "weight times lever arm" and "mass times acceleration." Heat is a measure of a quantity of energy, and temperature a measure of its intensity, so their juxtaposition seemed contrived; Clausius might as well have determined the ratio between the size of a wheel of cheese and a numerical rating of its sharpness. But despite its lack of obvious meaning, the equality of ratios held good, and furnished the second equation that was needed to complete the analysis of the energy budget of reversible engines.

From reversible engines Clausius went on to examine irreversible processes. He asked what happens to that same ratio—heat over temperature—when heat leaks from a hot cup of tea into the cool air and is dissipated in the atmosphere. The first law again guarantees that the total amount of energy is unchanged—it simply flows from the cup into the atmosphere. But since the temperature of the air is *lower* than the temperature of the cup, the heat to temperature ratio *increases*. If, say, a calorie of heat from a cup at a temperature of 330 kelvin meanders into a room at 300 kelvin, the ratio of heat to temperature increases from $1/330$ to $1/300$, or by about 10 percent.

Clausius examined as many such examples as he could. From chemistry, engineering, applied physics, and pure physics, from the behavior of gases, liquids, and solids, from the published literature, from his own experience, and from that of his colleagues on the Dream Team, he collected special cases of reversible and irreversible processes involving the transfer of heat. His analysis yielded the same result, until he became bold enough to announce it as a general rule: Heat over absolute temperature is constant in reversible processes, and increases in irreversible ones. This was what Clausius meant by "a different form of the second law" in the title of his paper, and now, finally, it was quantitative.

Driven by the principle of parsimony, Clausius did not stop there, but pressed on in search of the perfect expression. In 1865, fifteen years after his first foray into the study of heat, and two years before the birth of Maxwell's Demon, he collected his thoughts once more in a paper entitled "On various forms of the laws of thermodynamics that are convenient for applications." By then he was sure he had the essence right, but in order to put his mysterious ratio onto a more solid footing, he decided to name it. And as a German professor steeped in Greek and Latin philology, he had no trouble coining a suitable word.

Using the arbitrary abbreviation *S* for the ratio of heat to temperature, he explained:

> . . . since I think it is better to take the names of such quantities as these, which are important for science, from the ancient languages, so that they can be introduced without change into all the modern languages, I propose to name the magnitude S the *entropy* of the body, from the Greek word [for] a transformation. I have intentionally formed the word entropy so as to be as similar as possible to the word energy, since both these quantities, which are to be known by these names, are so nearly related to each other in their physical significance that a certain similarity in their names seemed to me advantageous.

In good German scholarly tradition, Clausius chose the name with meticulous care. He derived it from the Greek for "transformation" because, in contrast to energy, which describes the state of a physical system, entropy refers to changes in the status quo—through natural means or induced by artifice. Clausius's thoroughness paid off when his coinage was universally accepted by scientists and eventually began to spread into everyday discourse.

Archimedes's product "weight times lever arm" also merited a special name derived from a classical language: It came to be known as "torque," from the Latin for "twist." But although levers are much more common than heat engines, the word "torque" is locked away in physics books and the toolboxes of auto mechanics. The term "entropy," on the other hand, has flourished. As shorthand for the useful though meaningless ratio of heat to temperature, it continues to be part of the jargon of chemistry and engineering. Later, when it was finally revealed as a measure of molecular disorder, and of the amount of wasted energy in heat engines and other thermal processes, it would rise to a position rivaling that of energy in our understanding of the universe.

In terms of entropy, the second law took on the form it has in today's textbooks of thermodynamics: In reversible processes entropy is conserved; in irreversible processes it increases. But even that wasn't good enough for Clausius. To be sure, it was succinct, but besides brevity of expression, the principle of parsimony also requires breadth of applicability. So Clausius went for broke. He ended his 1865 paper with an irreducible statement of the laws of thermodynamics that is as awe-inspiring in concision as it is in scope:

1. The energy of the universe is constant.
2. The entropy of the universe tends toward a maximum.

By entropy Clausius meant only the enigmatic, macroscopic formula "heat over absolute temperature." The deeper meaning of entropy could not be discovered until the Demon led the way down into the microscopic realm of the atom.

8

RIVERS OF GOLD: A PARABLE

The universe is made of stories, not of atoms.

—MURIEL RUKEYSER, *THE SPEED OF DARKNESS*

A parable to illustrate the state of thermodynamics in the middle of the nineteenth century, after the formulation of its two laws by Rudolf Clausius, but before anyone understood the real meaning of entropy.

In a remote valley of the Hindu Kush mountain range, whose inhabitants speak an unknown dialect, shepherds stumble upon some peculiar springs. Chemical analysis reveals that the iridescent water bubbling out of the ground is uncommonly rich in dissolved gold. Geologists who quickly survey the area classify the springs in terms of their intensity, or concentration, of dissolved gold, measured in ounces per gallon of water. The local scientists, who insist on a jargon derived from the classical languages, call the concentration "temperature."

Soon channels are built to processing plants, where the metal is extracted and refined for export. Recovered gold is called "energy," and distinguished from gold in solution, which is called "heat," and at first nobody appreciates the quantitative relationship between the raw material and the finished product. (A long time ago a mad American count who had wandered into the area had suggested that they are, in fact, identical, but his ideas quickly passed into oblivion.)

A clever young engineer by the name of Sadi decides to embark on a thorough analysis of the extraction plants, which have become the engines of the local economy. He notices that all of them use ordinary water from local rivers for processing. For chemical reasons he does not fully understand, the plants cannot reduce the gold concentration of the valley water below that of the river water, so that the effluents of the facilities invariably bear residual traces of gold. This discovery suggests to him a way of measuring the caliber of a plant: The greater the difference between the concentrations of gold in the entering and discarded water, the better and more efficient the plant.

In this way Sadi can compare the efficiencies of extracting facilities without ever setting foot in them, just by knowing the gold concentrations of the water before and after processing. Sadi realizes that his comparisons show theoretical maximum efficiencies; gold wasted in the machinery of the plant always reduces the actual efficiency below his value.

A few years later, Roberto, a visionary physician, and Joulio, a down-to-earth brewer, independently discover that the total weight of gold—in solution as heat or in the extracted form of energy—is always conserved. The amount of gold that enters a plant must tally with the amounts extracted, lost, and discarded. This momentous discovery implies a definite relationship between dissolved and extracted gold, or heat and energy, and is celebrated as the first law of thermodynamics.

A little later Claudio, a professor of physics, rediscovers the wisdom of Sadi, and to explain it decides to focus on the water instead of the gold. He introduces the abstract concept of pure, gold-free water, something never found in nature, and coins the new word "entropy" to describe it. In an effort to demystify the term, he relates it to previously defined words. He points out that since concentration is defined as the weight of dissolved gold divided by the volume of pure water, the volume of water can be computed by simply dividing the weight of dissolved gold by the concentration. In the technical language of the land, this means that entropy is nothing but the ratio of heat to temperature.

By making the bold new assumption that the amount of water flowing through a channel into a plant equals the amount coming out, he can easily explain Sadi's results concerning the efficiencies of extraction plants. To the first law of thermodynamics, which guarantees the conservation of energy, he adds the second: Entropy is also conserved. With Claudio's two laws, scientists have a complete mathematical description of an extraction plant. Everything falls into place.

So Claudio presses on. He turns from the study of artificial channels and extraction plants to the natural flow of the rivers in his country. It has always been known that as you paddle down a river, the concentration of gold gradually diminishes. Heat flows naturally from hot places to cold places, the scientists say. How can this natural tendency toward lower temperatures be explained?

Claudio knows that the dissolved gold in the river cannot vanish into nothingness, so he concludes that natural rivers behave differently from artificial channels. Rivers continuously gather water from tributaries, hidden springs, and rain as they wind their way down to the ocean. This complex process cannot be controlled and renders the flow of a river irreversible, unlike the flow in a channel, which can easily be reversed. As the river swells, the gold it carries is diluted and the concentration decreases. Thus Claudio generalizes his second law: In artificial, reversible processes, entropy is conserved. In natural, irreversible processes, entropy increases.

When Western travelers finally penetrate to the capital of the land of the golden rivers, they are astonished to find, in the central library, a book laying out the complete theory of thermodynamics, with the same formulas they learned in their physics courses. They are puzzled, however, to find the book in the section on mining.

Clausius left the real question unanswered, the fundamental problem of thermodynamics in the Victorian age: What is this stuff called entropy? *Imagining it as water, as a material substance, especially one whose bulk grows with time, is just as false and misleading as interpreting heat in terms of the discredited caloric—or of gold. A better explanation was sorely needed, but it would entail a drastic narrowing of focus from the macroscopic to the microscopic. The secrets of heat reside in the least parts of matter, and to discover them requires a descent into the hidden world of molecules.*

9

A GAME OF BILLIARDS: THE STORY OF TEMPERATURE

Nature, it seems, is the popular name
for milliards and milliards and milliards
of particles playing their infinite game
of billiards and billiards and billiards.

—PIET HEIN

"I am never content until I have constructed a mechanical model of the subject I am studying. If I succeed in making one, I understand; otherwise I do not," claimed William Thomson, Lord Kelvin. Models, in the primary sense of the word, refer to manufactured objects, and as such play a role in all civilizations. Figurines of people and animals, together with the paraphernalia of their daily lives, fill the cabinets of anthropological museums. Closer to home, the popularity of miniature ships, soldiers, cars, trains, airplanes, and dinosaurs attests to the universal fascination of models. The scientific counterparts of these toys are the mechanical models of everything from the solar system to the structure of matter that reached the height of their popularity among physicists in the late nineteenth century. Kelvin's reliance on them was perhaps a bit extreme, but the double helix of DNA, which was discovered with the aid of a mechanical contraption resembling a Tinkertoy tower, reminds us that even in our time, models play an essential role in the quest to understand the world.

As physics became more sophisticated, the definition of the word "model" was generalized to include mathematical models, which don't necessarily have a tangible representation. A mathematical model is an idealized theoretical description that reproduces the basic features of a phenomenon, while leaving out its inessential complications. Galileo's law of free fall is a case in point: It describes the way a feather and a hammer fall in a vacuum, but, because it ignores air resistance, provides only a very rough approximation of what we actually see around us. The set of equations that is fed into a supercomputer to describe the interior of a star, similar in principle to Galileo's law though vastly more complex, is also a mathematical model.

The shorthand nature of mathematics often allows the same equation to describe different phenomena that display similar behavior. The swing of a pendulum, for example, obeys exactly the same equation as the vibration of a violin string. When that happens, the mathematical model reveals an analogy, a hidden likeness between superficially different sense experiences. It provides another step toward "a simplified and lucid image of the world," in Einstein's ringing phrase; in William Blake's more lyrical words it leads us "to see the world in a grain of sand." Early in this century, Niels Bohr managed to invert Blake's vision, and at the same time to surpass its vast compass, when he used the solar system as a model for the hydrogen atom.

The oldest and still the most powerfully suggestive model of physics is the atomic hypothesis, or "the atomic *fact*, or whatever you wish to call it," as Richard Feynman, who was more interested in meaning than in words, expressed it. The origin of the strange idea that the world, contrary to its solid and continuous appearance, consists of innumerable invisible kernels of matter imbedded in a void is unknown. For two millennia the atomic model of matter remained a mere speculation, weightless as a dream, barren of rigorous mathematical consequences that bear comparison with the empirical evidence of the senses.

But eventually contact with the real world was established. In the eighteenth century, the billiard ball model of a gas transformed the atomic theory into a useful model with predictive power. It was inevitable that scientists first used detailed atomic models to investigate the property of gases. The mechanical, optical, acoustic, and thermal properties of solids and liquids depend on the sizes and shapes of their constituent atoms and molecules, and also on the forces that bind them together. But as long as atoms were inaccessible to experimentation, all those detailed attributes of individual particles were unknown. In gases, on the other hand, atoms are so far apart that they don't influence each other very much, so their mutual forces, along with most of their other properties, are irrelevant: Only collisions with the walls of their container matter. The laws governing the behavior of gases resemble the formulas worked out by engineers to describe traffic on superhighways, which are unaffected by the sizes, shapes, and mechanical specifications of the cars. The numbers and speeds of the participants suffice to describe the behavior of traffic and of gases.

Many of the laws governing gases are very simple. Consider the relationship between pressure and volume of a fixed amount of gas: The harder you squeeze, the smaller it gets. When the plunger of a bicycle pump is pushed down, the diminishing volume of air in the barrel is accompanied by increasing pressure, until there is enough to overcome the pressure inside the tire and it inflates. Mathematically, this law is expressed as a simple inverse proportion between volume and pressure: As the volume diminishes, the pressure increases, and vice versa.

Daniel Bernoulli's *Hydrodynamica* (1738), a theoretical and practical treatise on fluids, did much to explain volume and pressure's inverse relationship. Bernoulli's family were Flemish, but had been driven from the Netherlands to Switzerland on account of their Protestantism. They were as accomplished and prolific in science and mathematics as the Bachs were in music. (So keen was the competition for fame in the gifted clan that Daniel's father, himself an eminent scientist, put some of the results of *Hydrodynamica* into a book entitled *Hydraulica*, which he predated to 1732 in order to steal the credit from his own son.)

Using air as an example, Daniel Bernoulli first assumed that all gases consist of minute particles, practically infinite in number, driven hither and thither with a very rapid motion and rarely colliding with one another. Since he imagined these molecules as hard and elastic, they can conveniently be thought of as miniature billiard balls.

In Bernoulli's model, pressure—defined as the force exerted by the gas on each square inch of wall—is caused by the relentless bombardment of the balls on the sides of the containing vessel. Now imagine a cubical box, one foot to the side, filled with air. If you double the length of each side without changing the number of balls or their speeds, the force exerted by each individual collision on a wall will remain undiminished. However, since it now takes a ball twice as long to cross the enlarged box to the other side and to return for another hit, there will only be half as many collisions per second, so the total force on the entire wall will be cut in two. At the same time the pressure, which is force per unit area, will suffer a further reduction by a factor of four on account of the fourfold increase in the area of each side (that is, each side is now four square feet instead of one square foot). Taken together, these two effects cause a drop in pressure by a factor of eight. But, as Bernoulli pointed out, that is exactly the increase in the volume when each side is doubled. While as pressure is an eighth of what it was, volume is eight times greater. Thus the model effortlessly accounts for the law that pressure and volume are inversely proportional to each other.

Bernoulli's billiard ball model of a gas has stood the test of time. Created long before the reality of atoms became a generally accepted fact, and before quantum mechanics and relativity irrevocably revised our way of thinking about them, its essential validity remains unchallenged. Furthermore, since the model can actually be realized in the teaching laboratory with Ping-Pong balls in order to illustrate the behavior of gases in a dramatic and convincing way, it serves as a conceptual stepping-stone from the world of immediate sense perceptions down to the inaccessible world of the atom.

But it accomplishes even more than that. Bernoulli managed to locate a little patch in the landscape of atoms where Newton's laws, which were derived from the observation of planets and billiard balls, happen to apply. We now know that when atoms coalesce into solids and liquids, and when they interact in chemical transformations, Newton's laws of motion are hopelessly inadequate. But by focusing on the singular problem of the behavior of gases, Bernoulli was able to bring the only truly fundamental laws of physics known in his time to bear on the basic building blocks of matter. In this way he rescued the atomic hypothesis from the realm of philosophy and for the first time installed it firmly in the province of physics. Bernoulli made it easier for people to believe in atoms by discovering a class of objects for which atoms were perfect models.

Bernoulli made one more suggestion concerning his billiard ball model:

> The elasticity of air is not only increased by compression but by *heat* supplied to it, and since it is admitted that heat may be considered as an increasing internal motion of the particles, it follows that . . . this indicates a more intense motion in the particles of air.

Here was a recipe for quantifying the idea that heat is motion—two generations before Count Rumford—but it came too early. Before Bernoulli's insight could bear fruit, the concept of the atom still had to endure a difficult period of gestation, much more had to be learned about the behavior of real gases, heat had to be rescued from the caloric theory, energy conservation had to be discovered, and the mechanical equivalent of heat had to be identified and then measured. Bernoulli's magnificently cogent model was forgotten, but when it was rediscovered and perfected more than a hundred years later, it was quickly adopted as the principal testing ground for the fledgling science of thermodynamics.

In our everyday experience of heat, air and other gases play a distinctly minor role: Hot tea and cold showers are more important than balloons and bicycle pumps for the formation of our ideas and intuitions about thermal behavior. Even the effects of air currents, the principal carriers of warmth and coolness outdoors, are often hidden under the influence of direct radiant heat from the sun, which is an altogether different phenomenon, unrelated to the heat carried by a gas. It is not surprising, therefore, that the first quantitative thermal experiments of Rumford and Joule concerned the heating of water, not of air. But once the simplicity of the gaseous state was appreciated, the students of heat turned eagerly to that less familiar form of matter. Joule himself lectured on the billiard ball model as early as 1847, and a decade later Clausius's poetically entitled "On the kind of motion we call heat" put it on a solid mathematical foundation.

The crux of the mathematical treatment of the billiard ball model is the realization that heat is just the sum total of the energies of motion, or kinetic energies, of the individual molecules. Thus, in the case of gases, at least, Rumford's oracular speculation about the true nature of heat was at last put into quantitative terms. At the same time another thermodynamic quantity was explained by the model. The absolute scale proposed by Kelvin, and used by Clausius in the definition of entropy, suggested a microscopic interpretation of temperature.

Hot air balloons are dramatic applications of the common observation that heating expands the volume of a gas. Conversely, cooling contracts gas, as flat tires in the winter remind us. This trend is a universal rule for all types of gas, and illuminates the fundamental meaning of the concept of temperature. Careful measurements on many gases have established the fact that as the temperature nears a chilly 273.15 degrees below zero on the Celsius scale, their volume collapses to zero. Since a negative volume is an impossibility, that magical −273.15 degrees is the lowest attainable temperature.

Thomson had the inspired idea of redefining the lowest conceivable temperature as zero, and calling that point the absolute zero. He pegged the melting point of ice at plus 273.15 degrees, and left the sizes of the intermediate temperature steps just as they were in the Celsius scale. The bottom of the resulting scheme—called the Kelvin scale in his honor—is not arbitrary, like the zero points of the older scales, but furnishes an anchor in a point of real physical significance. It corresponds to the convention of measuring the heights of mountains from sea level, which is a universal calibration point, rather than from some spot in a valley that is arbitrarily defined to be the foot of the mountain. By placing the definition of temperature on a more physical basis, the absolute scale led to a considerable simplification of the laws of thermodynamics and to a molecular interpretation of the concept of temperature. When measured on the Kelvin scale, temperature is nothing but the average energy of motion of a single billiard ball or molecule.

This molecular interpretation removes much of the mystery from the abstract concept of temperature. For one thing, it explains the nature of the absolute zero: When all motion ceases, the kinetic energy of the molecules is zero, and that's as far down as you can go. (Actually, only the random motion of the entire molecule ceases at absolute zero. The regular, periodic oscillations of electrons inside atoms, as well as those of protons and neutrons, and of quarks inside their nuclei, continue unabated.)

The molecular perspective also illuminates the difference between temperature and heat. A thermometer measures not the quantity but the *intensity* of warmth, meaning that temperature can be determined, like the sharpness of cheese, from a small sample of the material in question. Heat, in contrast, represents the totality of warmth. Like the total quantity of a hunk of cheese, it depends on the total mass of the gas. The billiard ball model reflects this difference. Indeed, it allows temperature to be measured by observation of the smallest imaginable sample of material—a single molecule. If the billiard ball model had been invented before the more conventional temperature scales, today's weather report might begin: "It's a balmy morning in New York. The average kinetic energy of air molecules is 5.1 sextillionths of a joule, but this afternoon it will rise to a blistering 5.3 sextillionths." And all of us would have an immediate visceral understanding of what those figures imply—because scientifically they mean exactly the same thing as the common designations of temperatures we are accustomed to.

In the middle of the nineteenth century, the billiard ball model accounted for all but one of the important properties of a gas. Heat, temperature, pressure, density, and weight were all interpreted in terms of hypothetical molecules and their assumed masses and hypothetical speeds. With one exception, every macroscopic measurement of a gas had a tactile, visualizable molecular interpretation, so the model succeeded brilliantly in allowing someone like Lord Kelvin to understand thermodynamics.

Only one challenge remained, the cardinal, unanswered question: How can the billiard ball model explain the new concept of entropy—Clausius's ratio of heat to temperature? By its very definition, the quantity was puzzling. Unlike torque, which relates position and weight of one object, and force, which links mass and acceleration of the same body, entropy refers to two very dissimilar entities. Heat is macroscopic and involves a collection of trillions of molecules, whereas temperature has turned out to be microscopic—a property of a single representative, average molecule. An understanding of entropy requires a more precise grasp of the relation of the parts to the whole, of how molecules combine to form the world of our senses.

10

THE DEVIL-ON-TWO-STICKS: CHANCE AND THE LOSS OF CERTAINTY

> *It is truth very certain that, when it is not in our power to determine what is true, we ought to follow what is most probable.*
>
> —RENÉ DESCARTES

When Isaac Newton was a schoolboy he built his own wooden model of the village windmill, but unlike Villard de Honnecourt or Robert Mayer, he had no unrealistic expectations for it. On the contrary, for indoor use he equipped it with a treadmill—so much for perpetual motion—and as a substitute source of power installed a mouse, whom he nicknamed "the miller." In later years, his love of constructing ingenious copies of the real world deepened and matured as mechanical models gave way to the more abstract, mathematical models that define the Newtonian philosophy. Two and a half centuries later young Albert Einstein, who was just as passionately interested in the way things work, looked upon the world in a more passive, contemplative manner. One of his most vivid childhood memories was the profound intellectual thrill he felt at the age of three or four when his father showed him a compass. Shivering with excitement, he watched the needle turn under the control of some unseen influence, and sensed that something deeply hidden had to lie behind the outward appearance of the device. He spent the rest of his life searching for the ultimate causes of the phenomena we experience, and never felt satisfied with mere mechanical or mathematical models of reality.

Compared with these two serious, introspective boys, James Clerk Maxwell was a much more playful and gregarious child. He too wondered about the workings of the world, but his father, to whom he grew uncommonly close after the death of his mother when he was eight, taught him to regard learning not as a solitary pursuit, but as a source of shared pleasure and excitement. Accordingly Maxwell approached science as teamwork, first with his family and later with an ever-growing circle of friends, colleagues, and students. As a child, his first reaction to an encounter with a new puzzle was to ask questions. "What's the go o' that? What does it do?" became a familiar refrain in the Maxwell manor house in the Highlands of Scotland from the time James began to talk. General or vague answers did not satisfy his thirst for understanding. "But what's the *particular* go of it?" he would insist, until someone—most often his father—took the trouble to give him a more specific explanation.

In 1843, when he was twelve, James was introduced to a game called the Devil-On-Two-Sticks. It consisted of a brightly painted wooden dowel in the shape of two cones joined at their tips to resemble an hourglass, and a string stretched between the ends of two sticks. By grasping the sticks at their free ends, and letting the dowel, balanced on its waist, roll down the length of the string several times, Master Maxwell was able to set it spinning like a yo-yo. Then he would fling the double cone into the air with a snap of the wrists and catch it up again on the string, which he could hold either in a slack loop or a tight line. During the Devil's short flight James, who quickly became an expert player, would jump over the string, turn on his heels, pick up a second Devil, or perform a variety of other nimble tricks of his own invention.

Long after Maxwell's death the game was named "diablo," or "diabolo," and became an international fad for children as well as adults, but in mid-century Scotland it was no more than a plain nursery toy. A superior imagination was required to appreciate its magical properties. Friends and neighbors must have smiled with indulgent puzzlement as they watched curly haired little James, dressed in the loose-fitting, sensible garments and square-toed shoes his father designed for him, pour his youthful energy into the curious game and, with an expression of intense concentration in his large brown eyes, attempt increasingly complex maneuvers. He became addicted to the "de'il," as it became known in the family's Highlands dialect, and tried to entice his friends to join in. Even much later, when he was a fellow of Trinity College at Cambridge—once Newton's college, and today Stephen Hawking's—he continued to play at the Devil in the gymnasium for relaxation and exercise.

Maxwell recognized—intuitively as a boy and rationally as a man—that a bout with the Devil was a contest between inanimate nature and the human will. With practice and patience he could learn to command the creature to do his bidding, recognizing full well that the dominion of the immutable laws of mechanics would always take precedence. These laws, which the boy could perceive only dimly, but which the man later put into the rigorous language of mathematics, seemed to endow the Devil with a will of its own. Like a gyroscope, it insisted on keeping the alignment of its axis: On order it would jump in all directions, but steadfastly refuse to change its orientation. In the air it would fall like a rock, but on the string it defied gravity like a graceful tightrope walker. All the while, young James was its master. By understanding its behavior, he could control it.

The Devil-On-Two-Sticks symbolizes Maxwell's approach to physics, just as the windmill represents Newton's and the enigmatic compass Einstein's. "At practical Mechanics I have been turning Devils of sorts," he wrote to a friend to explain what he was studying at the University of Edinburgh, comparing the solution of physics problems to the manipulation of the Devil. It was a matter of playing with the problem, tossing it around, looking at it from different points of view, poking at it and putting it through its paces until it became docile in his hands—until he became its master. He turned Devils of that sort all his life.

Maxwell's major contribution to science was the unification of the laws of electricity and magnetism into one coherent mathematical framework. He showed how the whole world of electromagnetic phenomena—from the force that pulls a nail toward a magnet, and the sparks that fly from a comb in the winter, to the most sophisticated applications of electric power and electronics, and even the sciences of optics and radiant heat—could be captured by four little formulas that came to be known as "Maxwell's equations." Through the power of this theory, Maxwell was able, for the first time, to compute the speed of light from basic principles, and to predict the existence of radio waves twenty years before their experimental discovery. As a grand synthesis, his theory is matched only by Newton's summing up of the entire span of mechanics in three simple laws of motion, and Einstein's reconciliation, by means of two seemingly innocuous principles of relativity, of Newton's mechanics with Maxwell's electrodynamics.

The method of the Devil-On-Two-Sticks, understanding through manipulation, lay at the base of Maxwell's accomplishment. Electricity, magnetism, light, and radiant heat are everyday phenomena, accessible to our senses and subject to experimental manipulation. More clearly than any of his contemporaries, Maxwell realized that theoretical physics must always remain rooted in the laboratory;

without fail his scientific arguments returned from their mathematical somersaults to the secure soil of immediate sense experiences.

But in spite of his spectacular successes, Maxwell eventually came to the limits of his method. When he began to think about thermodynamics, and the way it grows out of the behavior of atoms and molecules, he entered a realm beyond the direct experience of touch and sight, where the practical lessons of the Devil-On-Two-Sticks became inapplicable. If he was forced to abandon his childhood game, what would take its place?

He described the turning point with Victorian eloquence in his presidential address to the British Association for the Advancement of Science, meeting in Liverpool in 1870:

> I have been carried . . . into the sanctuary of minuteness and of power, where molecules obey the laws of their existence, clash together in fierce collision, or grapple in yet more fierce embrace, building up in secret the forms of visible things.

What a picturesque intuition Maxwell had about the world of atoms, decades before their existence was universally accepted, and a century before they became visible, and how painfully he must have felt that qualification "in secret." But he feared more than just his own failure to grasp "the particular go" of molecules; he worried about loss of contact with reality. "Who will lead me into that still more hidden and dimmer region where Thought weds Fact," his speech continued, "where the mental operation of the mathematician and the physical action of the molecules are seen in their true relation? Does not the way pass through the very den of the metaphysician, strewed with the remains of the former explorers and abhorred by every man of science?"

To discover the true relation between mathematics and experimental facts is the very essence of the physicist's enterprise, and to allow it to deteriorate into metaphysics would spell its demise.

Isaac Newton was able to hold on to a mechanical world picture throughout his life. His laws furnished a precise model of reality, and God, the heavenly miller who took the place of the light-footed miller of his childhood, kept the whole machine turning. Nor did Albert Einstein waver from his youthful insight: He found the ultimate reality of the material world in the invisible force fields of gravity and electromagnetism that suffuse and control it the way the earth's magnetic field governs the needle of a compass. Only Maxwell was compelled to change paradigms in midcareer. Characteristically, he did not ask which way to turn, but rather *who* would guide him in the right direction. People always mattered more to him than things.

In discarding the Devil-On-Two-Sticks as his familiar, Maxwell reflected a sea change in physics. Since antiquity, mechanics, the science of motion, had been applied to objects that can be seen and, in many cases, manipulated—objects whose reality was not open to question. The essence of Newton's method was to identify the forces acting on an object, to predict the course of its motion with mathematical rigor, and then to check that prediction against direct observation. The underlying assumption was that mechanical systems behave in definitely predictable ways. However complex and incomprehensible the gyrations of the Devil might seem, they were always certain.

When it came to molecules, the rules of the game suddenly changed. For one thing, molecules were of dubious reality, but that wasn't what bothered Maxwell; he didn't mind going out on a limb, and was quite prepared to throw out the billiard ball model if molecules turned out to be mere figments of the imagination of Leucippus, Democritus, Lucretius, and their followers. What troubled him was the fact that molecules were too small to be seen, too numerous to be accounted for individually, and too quick to be captured. In the "sanctuary of minuteness and power" of molecules, he no longer had the control he had become accustomed to from childhood. He had lost the luxury of certainty.

But if Maxwell couldn't work with full knowledge, he would make do with partial information; if certainty was beyond his reach, he would settle for the next best thing—and so he turned to probability. Although it was born of necessity, and attributable to the unwieldy number of particles in the atomic world, Maxwell's conversion turned out to be the crucial event on the road toward a correct understanding of entropy. Maxwell didn't achieve that goal himself, but without his discovery of statistics as a new tool in the arsenal of physicists, entropy would have remained the useful though ultimately meaningless ratio of heat over temperature.

Maxwell's reputation as the greatest theoretical physicist between Newton and Einstein rests only in part on his pioneering technical contributions to the new sciences of electro- and thermodynamics. On a deeper, more philosophical level, he was a radical reformer who cleared the way for the scientific revolutions of the twentieth century. By replacing the Newtonian action-at-a-distance formulation of electromagnetic forces by fields, he established the conceptual framework for the general theory of relativity—the theory of the gravitational field. And then, by switching from certainty to probability as a basis for thermodynamics, he put into place an essential cornerstone of quantum theory—the modern description of the interior of an atom.

The developments that led to the quantum theory resemble the steps leading to the modern theory of thermodynamics. In the first quarter of the twentieth century, physicists crossed the outer skin of atoms and began to investigate their hidden interior landscape, where electrons swarm around a tiny, central nucleus. They invented increasingly complicated models of atomic structure, using as a guide Newton's mechanics, which had proven to be so successful in the description of the solar system. But where Maxwell had been stymied by unmanageably *large* numbers, the atomic physicists suffered from a lack of experimental access to the interior of the atom because it is so *small*. In the end they had to admit defeat. Their elaborate conjectural models of the interiors of atoms simply didn't work.

In 1925, the young German physicist Werner Heisenberg, frustrated by lack of progress in understanding the atom, swept the slate clean. Instead of continuing to invent electronic orbits that he had no hope of verifying, he decided to work with the precise but admittedly partial information that he could count on, such as the colors of light emitted by atoms and the amounts of energy exchanged in atomic encounters. At this point he consciously gave up the illusory certainty concerning the motion of electrons in atoms that the models had taken for granted. Instead, he stipulated only that the electrons occupy different regions in the vicinity of the atomic nucleus with different probabilities—but didn't attempt to pin them down more precisely. It was a revolutionary way to think about the interior of the atom, but it yielded spectacularly accurate results in the form of predictions about atomic behavior. In this way Heisenberg and his fellow quantum mechanics unexpectedly introduced the notion of probability into the realm of atomic physics, just as Maxwell had brought it into the science of heat.

In the end, the quantum mechanical uncertainty of positions and speeds of particles turned out to be more fundamental than that of thermodynamics. Where Maxwell appealed to probability because he didn't happen to know the details of the motions of invisible molecules for *practical* reasons, quantum mechanics insists that we cannot know the exact whereabouts of a subatomic particle for reasons of *principle.* Even the Demon is powerless inside an atom. But aside from this fundamental difference, the fact remains that without Maxwell's introduction of the mathematics of chance into the description of nature, the quantum revolution could not have come about.

11

HEADS AND TAILS: THE LAWS OF PROBABILITY

There are few laws more precise than those of perfect molecular chaos.

—GEORGE PORTER

When James Clerk Maxwell turned to probability theory for help in understanding heat, he had to overcome not only two millennia of scientific tradition, but also his own religious scruples. At the age of nineteen he had written to his lifelong friend Lewis Campbell: "Logic is conversant at present only with things either certain, impossible, or *entirely* doubtful, none of which (fortunately) we have to reason with. Therefore the true logic of this world is the Calculus of Probabilities . . . This branch of Math, which is generally thought to favour gambling, dicing, and wagering, and therefore highly immoral, is the only 'Mathematics for Practical Men,' as we ought to be." By "entirely doubtful" he meant a situation about which nothing whatever is known, a state of affairs more extreme even than the uncertainty that hangs over molecules. Like his father, Maxwell was always a thoroughly practical man, so when he found himself in an intellectual corner, he turned without flinching to the morally repugnant, though mathematically impeccable, calculus of chance.

The immediate cause of Maxwell's dilemma was his

vivid imagination. To him, molecules were more real than to his contemporaries—in his mind's eye he could see them whizzing about like hailstones in a tempest. Many years later, Ernest Rutherford, the discoverer of the atomic nucleus, had a similarly graphic vision of atoms and molecules at a time when most of the world still regarded them as convenient fictions. "Jolly little beggars, so real that I can almost see them," Rutherford called them when someone questioned their reality.

To Maxwell's contemporaries, the billiard ball model of a gas was a useful mathematical device more than a faithful description of reality, so they felt perfectly free to modify it in any way they chose. Joule, for example, simplified it to the point of ludicrousness. In a public lecture he put it as follows: "Let us suppose an envelope of the size and shape of a cubic foot to be filled with hydrogen gas. Further, let us suppose the above quantity to be divided into three small elastic particles, each weighing 12.309 grams." A gas consisting of three monster molecules, each weighing as much as a walnut and moving along one of the three axes of a Cartesian coordinate system, would certainly simplify the calculation—but it had little to do with reality. The extreme precision of the numbers, together with the primitive nature of the accompanying mathematics, was characteristic of Joule's approach to physics, although it must be added in fairness that he quickly went on to allow his listeners to mentally subdivide his three particles into as many pieces as they wished.

Clausius, who was a professional theoretician and a much better mathematician, used a different device to overcome the intractability of the billiard ball model. Rather than lumping all the molecules into one, or three, as Joule had done, he worked with average values—average speed, average energy, average time interval between collisions, average distance between molecules. Had he been a sociologist, he would have described the individual behavior of Mr. and Mrs. Average, and imputed it to the entire population. Even Clausius's method didn't satisfy Maxwell. In his imagination he could see the hurly-burly of a real gas in which some particles acquire excessive speeds through collisions, while others temporarily give up most of their energy and consequently slow down. In between, Maxwell knew, all possible speeds and energies must be represented.

The motion of molecules in a gas is more complicated than Joule and Clausius assumed. In fact, the Belgian chemist Jan van Helmont had chosen well when he derived the word "gas" from the Greek for "chaos." If we want to understand a gas fully, Maxwell maintained, we must come to grips not only with the average behavior of molecules, but with all their wild cavortings. But how to penetrate that veil of secrecy under which nature chose to hide them?

The calculus of probability to which Maxwell turned is a tool of astonishing power and acuity. Its results are more easily accessible than those of other fundamental laws of nature. To witness the relentless way in which probability imposes its elegant patterns on random events, and molds formless mounds of data into beautiful shapes, you need nothing more than a penny and a lot of patience, or patience and a lot of pennies.

The outcome of the toss of a coin is, in Maxwell's phrase, entirely doubtful—we know nothing about it in advance. We are accustomed to the proposition that the probability of throwing heads is one half, but *applied to a single throw,* that statement is meaningless. It is unable to predict anything, it doesn't describe anything, and it certainly cannot be verified; after the coin has landed, the outcome of the experiment has become a certainty, so the word "probability" ceases to apply. For one single throw, indeed for every single throw, probability defies definition.

Only with repeated throws does its meaning gradually emerge. A hundred flips will yield approximately fifty heads, and as the number of tosses increases, the fraction of heads inexorably approaches closer and closer to one half. To witness the emergence of a more intriguing pattern, I solicited the help of my daughters, Madelynn and Lili, ages ten and eight. I gave them ten pennies and a yogurt cup, and showed them how to plot the results of successive throws of all ten using a bar chart of frequency versus head count—when four heads came up, they were to add a square to the top of the growing column of fours, and similarly for five heads, six heads, and so on, from zero to ten. After a dozen tosses their diagram, conscientiously constructed under my supervision, and meticulously shaded in Lili's heavy hand, looked bunched toward the middle, but otherwise shapeless. After twenty tries, the two ends of the diagram, where the zeros, ones, nines, and tens were recorded, were still conspicuously empty. After forty throws we called a huddle.

The girls were very disappointed by the absence of unusual events at the ends of the distribution. I tried to console them by explaining that the chance of throwing ten heads, or none, is about one in a thousand, so that at this stage of the game we really shouldn't expect any. But I pointed out something more disturbing: Where the sixes were recorded was a hole. In forty tries there had been only two sixes, and we were at a loss to explain the shortage. I encouraged them to keep up the good work, and went back to my study.

After three minutes a shout went up: "Daddy, come quickly, we got something!" Display of the evidence: The ten pennies all showed tails; skepticism on my part, outraged protestations of honesty on theirs, general glee all around. The left-hand edge of the graph sprouted a box at zero—an outlying island of improbability. But at the sixes the mountain was still split by an ominous cleavage. "Don't worry," I said, "it will fill up. This is exactly the kind of fluke that keeps gamblers hoping and gambling. But casino owners always win. Wait and see."

For half an hour I heard nothing but groans and muffled shouts. Between throws forty and fifty, a remarkable run of three sixes in a row went a long way toward filling up the hole. After 120 tosses, the girls came in to show me the plot. Besides the famous zero it sported a solitary nine, and the gash in the middle had disappeared—healed over by the law of averages.

Madelynn and Lili made two hundred throws that day before they gave up, and the picture they produced is a little miracle. Its outline is a bit jagged, but the pronounced maximum in the middle, the pleasing symmetry, and the famous bell shape of the curve are unmistakable. In red I drew the theoretical prediction derived from the calculus of probability over the girls' experimental points, and the agreement was satisfactory. What's more, I knew that if they had had the stamina to go on to a thousand throws or beyond, and if they had used a hundred pennies instead of ten, their laboriously collected data would have filled the red outline as smoothly as a hand fills a glove. I also knew that I could have programmed my computer to do all this, and more, in a fraction of a second. But I would have failed to make contact with the firm soil of experience, and worse, I would have missed the pleasure of watching two little girls confronting a law of nature in wide-eyed wonder.

The way the bell curve emerges is nothing short of magical. That the distribution should be symmetrical about the middle is not surprising—after all, what's good for heads should go for tails as well. That the graph should fall off toward the ends also agrees with intuition. We have enough daily experience with chance events and news accounts of lottery results to marvel at a throw of ten tails, or even at nine out of ten. The maximum in the middle is a reflection of the large number of configurations in which five heads can be realized. In fact, five heads can be picked out of ten coins in 252 distinct ways, whereas nine heads can be found in only ten different ways (corresponding to the ten ways of picking out the lone tail). Ten heads correspond to a single, unique configuration. The most probable outcome of a throw, in other words, is the one which can be achieved in the greatest variety of ways. Nature prefers variety.

The general features of the bell curve make intuitive sense, but why it should have precisely the shape it does, which I can confidently predict with the help of a few lines of grade school arithmetic, and not some other shape—a little wider in the hip, say, or more pointed, like a witch's hat—that remains a mystery. What power guides the pennies, whose individual motions are completely random, to fall in such a predictable way? What intelligence orders them to fill in the valley of the sixes instead of digging it deeper? How does this exquisite order emerge from the chaos of pennies tumbling pell-mell out of a cup?

The literature on the calculus of probabilities is vast, but it is powerless to dispel the magic. Since the eighteenth century mathematicians have derived increasingly general propositions, collectively known as "central limit theorems," to explain the astonishing universality of the bell curve, but its reappearance, over and over again, in the most diverse circumstances, still excites our wonder. Every now and then we must remind ourselves that the toss of a penny, the cooling of a cup of tea, and the twinkle in the eye of a child are mundane miracles we would do well to celebrate as undeserved gifts.

Maxwell applied the law of averages to the billiard ball model of a gas. Using simple arguments similar to those employed by mathematicians to derive the rules governing the tosses of coins and dice, he worked out a formula for the speeds of molecules. When it is plotted on a graph it looks very much like the bell curve my daughters discovered—the number of molecules with very low speeds is small, then the count rises to a maximum around the average speed used by Clausius, and finally falls off again, almost symmetrically, until it becomes vanishingly small at extremely high speeds. (The symmetry about the average is spoiled a little by the fact that the curve ends on the left at zero speed, but extends on the right all the way up to infinity—at least in principle, since the natural speed limit for particles, set by the laws of relativity, is not respected in this approximate calculation.)

Unlike Madelynn and Lili, Maxwell had no direct way of verifying his theory. Instead, he drew a number of far-reaching, indirect inferences from it which were in due course checked experimentally. Thus it came about that Maxwell's theory of molecular speeds was accepted as correct long before twentieth-century technology allowed its direct corroboration. But not everyone appreciated Maxwell's theory. Clausius, for example, could never fully accept Maxwell's use of probabilistic arguments. Perhaps Clausius was too much the German professor, and lacked the playfulness that Maxwell had acquired as a child and never lost as a man. Games of chance were immoral according to Maxwell's strict Victorian code, but that didn't prevent him from trying to figure out the go of them, any more than he had shrunk from playing with the very Devil. But who would lead him into that dim region of molecular chaos where entropy arises?

12

THE MECHANICAL DEMON

> *The conception of the "sorting demon" is purely mechanical . . . It was not invented to help us deal with questions regarding the influence of life and of mind on the motions of matter.*
>
> —WILLIAM THOMSON, LORD KELVIN

In December 1867, James Clerk Maxwell wrote to his friend Peter Guthrie Tait at the University of Edinburgh about the behavior of gases. Basing his reasoning on the billiard ball model, he tried to predict what would happen if he could control the molecules one by one. Since he obviously couldn't, he invented a servant to do it for him. Thus Maxwell's Demon, a "very observant and neat-fingered being" who could see and manipulate individual molecules, was born.

Maxwell imagined him as the gatekeeper at a little door in the wall between two boxes full of gas that are initially at one common temperature. According to the microscopic interpretation of temperature, both boxes contain balls with a wide spread of speeds, even though the average speeds on opposite sides of the wall are equal. The Demon allows only exceptionally fast balls to pass from the right into the left chamber, stopping all others. (To keep the numbers of particles in the two boxes equal, he might also be instructed to allow exceptionally slow molecules to pass in the opposite direction.) The result would evidently be that the left box gets warmer, and the right one cooler. Heat would begin to flow without a difference in temperature, like water flowing along a level trough, and then continue on uphill—a clear violation of the second law of thermodynamics.

Of course, Maxwell did not believe in the existence of microscopic sorting sprites. His fanciful creation of the Demon was not the conjuring of a spirit but the supposition of a mechanism that could test his theories.

As a research tool the Demon quickly proved his worth: As soon as he was born he began to raise difficult and significant questions. To begin with, one wonders whether such a mechanism can actually be constructed. If the answer is no, why not? On the other hand, if the answer is yes, what would that imply about the second law? Would the Demon invalidate the entropy principle as a universal law of nature? Maxwell did not know.

The first step toward solving any problem is to formulate a crisp statement of the question. Since the original conception of the Demon was too charmingly vague to be analyzed systematically, Maxwell undertook to describe him in greater detail, and came up with what might be called "the mechanical Demon." Once this was done, he probed the idea with every theoretical argument he could muster. If he could find a flaw in his own reasoning, the Demon would be "dead," a possible objection to the second law would be removed, and its stature as a robust law of nature would be enhanced. On the other hand, if he could find nothing wrong with his Demon, he would have to concede that violations of the second law are, *in principle,* possible, if only you could create the right sort of mechanism, the equivalent of a perpetual motion machine. This would, in turn, imply that the second law is very different from the first, to which no exception has ever been discovered.

In pondering the Demon, one might worry that he consumes energy in opening and closing his little door. If that were true, he would merely act as an ordinary refrigerator, which uses up electrical energy to push heat uphill, out of the icebox and into the kitchen. But this particular worry can be easily dismissed. One must simply imagine that the door and the Demon are so small, and so exquisitely constructed, that the energy they require for their operation is negligible in comparison with the energy carried by any single molecule.

More troubling are the questions about the details of how the Demon goes about his work. How does he measure velocities? Does the process of measurement consume energy? How exactly does he come to a decision about a particular molecule? Do his judgments and choices exact an expenditure or degradation of energy, or are they "free"?

In order to get around such subtleties, Maxwell followed a time-honored tradition of physics: He proceeded to simplify his assumptions. He proposed an alternative arrangement in which the two boxes again start at equal temperatures and pressures, but he instructed his imaginary Demon to ignore their speeds. Instead, they were to allow molecules to pass from the right to the left, but not in the opposite direction. In this way no complex measurements or decisions needed to be made, yet the pressure of the gas would gradually rise in the left box and diminish on the right. After a while, the resulting difference in pressure could be used to drive a turbine, and work could be extracted from the two boxes without discarding waste heat: another, slightly different violation of the second law.

Maxwell emphasized that the simplified Demon was no more sentient than a machine. "This reduces the Demon to a valve. As such value him," he urged Tait, but neither man knew what to make of that valve. Maxwell would go to his grave without knowing whether his mechanical Demon could or could not violate the second law.

Credit—or blame—for the demise of the mechanical Demon belongs in large part to Albert Einstein, even though he did not personally participate in the goblin's death. One of the three heroic deeds of his miracle year of 1905 had been the explanation of Brownian motion, the microscopic, random dance of pollen particles in drops of still liquid and of smoke particles suspended in the air. Einstein imagined the particles as enormous boulders buffeted continuously from all directions by an immense swarm of hailstones representing individual atoms. On average, the effect of the hail is the same in all directions, but from time to time a statistical fluctuation occurs, like the valley in my daughters' bell curve. If, for a short time, a larger than average number of hailstorms should happen to bombard the boulder from the left, it would move to the right a little bit before settling down again. Instead of ignoring statistical fluctuations, Einstein proposed that they are responsible for Brownian motion, and reduced their effect to a simple mathematical equation. When experimentalists later compared Einstein's predictions with actual observations, they provided the first convincing, albeit indirect, evidence that atoms are real, physical objects—not just convenient fictions for simplifying the bookkeeping of chemical reactions or populating picturesque models of gases.

Marian von Smoluchowski, a brilliant Polish physicist who had worked with Lord Kelvin, independently followed a line of thinking very close to Einstein's. In fact, he had explained Brownian motion before 1905, but because he lacked experimental verification of his theory, had allowed Einstein to beat him to the punch. A little later, in 1912, Smoluchowski invoked Brownian motion to kill off the mechanical Demon.

Specifically, he imagined the door between the two gas-filled boxes to be a simple flap free to open toward the left only. Gone were the large brain and the deft fingers—nothing remained but a valve. Could such a device cause a rise in pressure on the left, and a corresponding drop on the right?

Smoluchowski argued that at first the flap might indeed admit a few extra molecules into the left-hand box, but that it would soon feel the effects any large foreign object feels in a gas: It would heat up, and then begin to jiggle with Brownian motion. As a result of its uncontrolled and random openings it would allow molecules to pass through in the wrong direction, thus restoring equilibrium.

But no theoretical argument unsupported by experimental evidence is ever completely satisfying, so Smoluchowski's cogent arguments did not end the controversy. Half a century later, when Richard Feynman, who liked to do things in his own way, set out to write his monumental textbook *The Feynman Lectures on Physics*, he invented a different version of the automated Demon. Feynman's machine is a minuscule ratchet-and-pawl mechanism that allows a small weight hung from a string on a pulley to rise but not to fall. The ratchet wheel is attached to a tiny windmill in a vessel full of hot gas. Left to itself, the windmill would turn this way and that under the constant bombardment of gas molecules. However, the pawl—the little hinged tongue that keeps the ratchet wheel from slipping—should ensure that the weight can only rise. But alas, it won't. Feynman showed that the pawl too warms up and bounces randomly, thereby allowing the weight to fall at random intervals. Careful calculation proved that this happens just often enough to save the laws of thermodynamics. Feynman wrote the blunt obituary of the dumb version of the Demon: "[It] must heat up. . . . Soon it is shaking from Brownian motion so much it cannot tell whether it is coming or going, much less whether the molecules are coming or going, so it does not work."

Experimental verification of this clear and unambiguous theoretical prediction still awaits the development of nanotechnology delicate enough to harness Brownian motion—

but modern computers can take us at least part of the way there. As recently as 1992, Wojciech Żurek, a physicist at the Los Alamos National Laboratory in New Mexico, together with a colleague from MIT, constructed a computer simulation of two boxes connected via one of Smoluchowski's trapdoors, with five hundred simulated billiard balls representing the gas. The result of this numerical experiment was a corroboration of Smoluchowski's and Feynman's calculations: The door soon shook with Brownian motion, and as a result, the second law held firm. The Demon—at least in his mechanical incarnation—has been well and truly thrown out the window.

The fact that as late as 1992 physicists still thought it worth their while to verify Smoluchowski's analysis of eighty years ago vividly illustrates their nervousness when dealing with the Demon. It has taken a very long time indeed for them to come to terms with Maxwell's simple little valve! But today we can confidently assert that the mechanical Demon is fully understood, and that he cannot violate the laws of thermodynamics.

This conclusion in no way implies that the Demon's life was a failure. On the contrary, the thoughts and debates on the subject of Brownian motion that he stimulated have enhanced our appreciation of that wonderful window through which Nature makes the motion of atoms visible to our poor human eyes. And, of course, our inability to see atoms was exactly the problem that frustrated Maxwell when he tried to understand the behavior of gases. In that sense, then, both the creator and his creature have succeeded splendidly.

Nor is the Demon really dead. Whenever he was tossed out a window, he has always managed to climb back in through another one in a new guise. At the end of a second paper on this problem in 1914, Smoluchowski announced the demise of the mechanical Demon, but added: "Such a device might, perhaps, function regularly if it were appropriately operated by intelligent beings." With that surmise he renewed the Demon's lease on life.

Maxwell invented his neat-fingered being to help him discover the microscopic meaning of entropy, but the Demon failed to guide him to that elusive goal. It took a human being, an Austrian physicist who turned out to be all too human in the end, to solve the riddle.

13

BOLTZMANN'S UNIVERSE: THE NATURE OF ENTROPY

The future belongs to those who can manipulate entropy; those who understand but energy will be only accountants. . . . The early industrial revolution involved energy, but the automatic factory of the future is an entropy revolution.

—FREDERIC KEFFER

On a fine summer evening in 1905, a corpulent gentleman dressed in a tight, rumpled three-piece suit, a starched white collar, and a pair of dilapidated shoes came bustling into the Vienna railroad station. He seemed to be in a hurry. The pugnacious thrust of his massive head, topped by a thick mane of curly black hair, contrasted oddly with the soft roundness of his bulging shoulders. His mouth was hidden under a straggly black beard in which two gray streaks were beginning to betray his age of sixty-one years. The man might have passed for a retired assistant clerk had it not been for the stubborn energy that radiated from his piercing dark eyes behind tiny rimless glasses whose frames seemed to bite painfully into his fleshy nose. His gaze was steady and profoundly serious, cutting mercilessly through the appearances of the world to its essences—yet at the same time it was marked by an ineffable melancholy.

Ludwig Eduard Boltzmann, professor of theoretical physics at the University of Vienna, world-renowned discoverer of the nature of entropy, later described the occasion:

> On the eighth of June I attended as usual the Thursday meeting of the Vienna Academy of Sciences. As I left the meeting a colleague noticed that I turned toward the Stubenring instead of toward Bäckerstrasse, as was my usual habit, and he asked where I was going. San Francisco, I replied laconically. In the restaurant of the Northwestern station I consumed suckling pig with sauerkraut and potatoes and drank a few glasses of beer. My memory for numbers, usually quite good, is always bad at recalling the number of glasses of beer I drank.

Boltzmann's reply to his colleague was accurate, but in our jet-propelled age it is difficult to appreciate the humor of the manner in which he delivered it. In 1905 a trip from Vienna to California, more than a quarter of the way around the globe, was a once-in-a-lifetime adventure, not the stroll around the block that Boltzmann made it out to be. More important, that factual answer, and the minute record of his menu on the eve of his voyage to the New World, reflect an attitude that is characteristic of Boltzmann, and indeed of science in general.

Unlike philosophy and religion, which seek absolute truths and find instances of them in the particulars of the world, science proceeds in the opposite direction, from the particular to the universal, as Boltzmann once explained in the course of a lecture on thermodynamics:

> The scientist asks not what are currently the most important questions, but 'which are at present solvable?' or sometimes merely 'in which can we make some small but genuine advance?' As long as the alchemists merely sought the philosophers' stone and aimed at finding the art of making gold, all their endeavors were fruitless; it was only when people restricted themselves to seemingly less valuable questions that they created chemistry. Thus natural science appears completely to lose

> from sight the large and general questions; but all the more splendid is the success when, groping in the thicket of special questions, we suddenly find a small opening that allows a hitherto undreamt of outlook on the whole.

By keeping his eyes firmly focused on immediate, attainable goals, like turning the corner toward the Stubenring and eating suckling pig in the Northwestern station, Boltzmann managed to reach the promised land of Eldorado—his private name for California.

The purpose of the trip was to teach a course of lectures in the summer session at the University of California at Berkeley. The obstacles Boltzmann had to overcome to reach his destination went far beyond the rigors of the ocean crossing and the four days and four nights of "shaking and rattling" on the Southern Pacific railroad. By this time of his life he was constantly plagued by a variety of illnesses: deteriorating eyesight, asthma attacks, angina, and migraines. Worse, he suffered from deep depressions that periodically carried him off into his own private hell, and had led to a suicide attempt a few years earlier. Not long before the California adventure, Boltzmann's wife, Henriette, had lamented to their daughter: "Father gets worse every day. I have lost my confidence in the future."

All these burdens were compounded by what Boltzmann called "a little stage fright before giving a first lecture," but which his biographers described as a more serious affliction. Boltzmann was a brilliant, even charismatic lecturer, at the cost of severe personal suffering. In preparation for California he had gone to the additional trouble of taking lessons to improve his spoken English, so one can imagine his anxiety at the prospect of lecturing in unfamiliar surroundings in a strange tongue.

But none of these impediments were enough to stop Boltzmann. Inside that ungainly body he was a tough, resilient fighter, prepared to rise to any challenge. Throughout his career, and especially in its latter part, Boltzmann had defended unpopular scientific ideas against formidable opposition and kept the faith; a mere jaunt around the globe to another world was a trifle by comparison.

The controversy in which Boltzmann was embroiled concerned the reality of atoms. Like Maxwell, he believed fervently that the billiard ball model and its refinements were more than convenient metaphors. He was convinced they were faithful representations of the real world. Since atoms stubbornly refused to reveal themselves, a growing number of physicists, including some whose scientific accomplishments Boltzmann freely acknowledged, and others whom he counted among his best friends, turned against a realistic interpretation of atoms. "I am conscious of being only an individual struggling weakly against the stream of time," he wrote sadly in 1898, but he never ceased fighting for the atomic theory. The very purpose of his life, he said, was "the development of theory, for whose glory no sacrifice is too great for me." Not even a voyage to Eldorado.

The small, solvable problem that had started Boltzmann on his brilliant career was a refinement of Maxwell's theory of the speeds of molecules in a gas. When Boltzmann was a student, his teacher had handed him Maxwell's papers, "and an English grammar in addition, since at that time I did not understand one word of English." An immediate, deep intellectual kinship had formed across the language barrier. Boltzmann, a talented pianist who was cajoled into playing Schubert for William Randolph Hearst's mother in California, recognized the mathematical styles of the masters of theoretical physics as easily as music lovers distinguish the styles of the great composers. Maxwell impressed him above all others. "Who doesn't know his dynamic theory of gases?" Boltzmann wrote, comparing Maxwell's work to a symphony:

> At first the variations of the velocity are developed majestically; then the equations of state come in on one side and the equations of central motion on the other; ever higher the chaos of equations surges until suddenly . . . a wild figure in the basses, which earlier undermined everything else, suddenly falls silent. As with a magic stroke everything that earlier seemed intractable falls into place. There is no time now to say why this or that substitution is introduced; whoever doesn't feel it should put the book away; Maxwell is no composer of program music who has to add an explanation to the notes.

Boltzmann recognized Maxwell's gas theory as a magnificent composition, and his own variations rendered it even more splendid.

The original billiard ball model had been developed with the simplifying assumption that the molecules are too far apart to interact with each other. For this reason, the only energy any ball carried was its kinetic energy of motion, which, in turn, characterized the temperature of the gas. But in the real world, molecules may attract each other and form pairs, which can rotate and vibrate in addition to flitting around in tight embrace, and which may even store potential energy like tiny compressed springs. Boltzmann set out to generalize Maxwell's theory to account for these additional kinds of energy, and in 1868, at the age of twenty-four, when Maxwell was still alive, succeeded. The formula he derived for the speeds of molecules in a gas is the celebrated "Maxwell-Boltzmann distribution."

Undoubtedly Boltzmann's Berkeley lectures dwelled on this "small but genuine advance" that extended the range of the billiard ball model to a host of new phenomena. But Boltzmann had other things on his mind in far-off California besides billiards. Perhaps the most exasperating discovery he made was that drinking was strictly prohibited in the Berkeley of 1905, a devastating problem because he also found that the local water didn't agree with his Austrian stomach. When he overcame his misgivings and divulged his predicament to a colleague, he was directed to a wine dealer in Oakland. "I managed to smuggle a whole battery of wine bottles across the city line, and I soon became quite familiar with the route to Oakland," he reported. "But I had to drink my postprandial glass of wine surreptitiously, so that I myself almost got to feeling I was indulging a vice." And with his customary insight Boltzmann went on to foresee the troubles that many others would face in the years to come: "The temperance movement is well on its way to giving the world a new species of hypocrisy. Surely we have enough already."

As Boltzmann was writing these prophetic words, Prohibition was still fifteen years away, but a greater revolution was at hand in his own discipline of physics. In the same year, Einstein's three papers on special relativity, Brownian motion, and light quanta had signaled the end of the Newtonian hegemony. Classical, Newtonian physics concerned phenomena at the ordinary human scale, and comprised mechanics, Maxwell's electromagnetism (including optics), and thermodynamics. The new century would bring advances into the non-Newtonian realms of the very fast, where special relativity applies, out into the cosmos, where general relativity takes over, and into the microworld of atoms, where the quantum theory reigns. It was fitting that Boltzmann, the last of the great classical physicists, came to California in 1905. His visit at the turn of the century symbolized the transition not only from the old to the new physics, but also from the Old to the New World, where many of the great discoveries of the twentieth century would be made.

For Boltzmann found the grail that had eluded Maxwell and his Demon: a microscopic interpretation of entropy. Entropy, defined rather enigmatically by Rudolf Clausius as the ratio of heat to temperature, is a macroscopic quantity: It is determined by measurements with ordinary laboratory instruments like thermometers, scales, and pressure gauges. In this view, it does not refer to molecules or their properties. Like volume, weight, and energy, it is additive: If you connect two identical vessels of gas at the same temperature, the new, enlarged system will have twice the volume, twice the weight, twice the energy, and twice the entropy of each individual vessel.

Boltzmann thought about the operation of joining those two vessels, and as a disciple of Maxwell's he imagined it in molecular terms. Since no convincing mechanical interpretation of entropy had yet been put forward, he sought its meaning in probability instead. From his experience with the bell curves of coin tosses and molecular speeds, he knew that blind chance can carve precise laws out of chaos, so he set out to search for a way to connect the inexorable growth of entropy with nature's pronounced preference for randomness and variety. But in his search for the relationship between entropy and probability he immediately ran into a fundamental problem.

Like Joule, Boltzmann might have imagined that the two vessels each contained one single gas molecule, but instead of assigning a unique speed to each, the way Joule had done, he would have been faithful to Maxwell in allowing for a spread of speeds. To keep matters simple, imagine that there are only five different speeds—one at the average value characteristic of the vessel's temperature, two a little lower, and two a bit higher to make a crude bell curve consisting of five points. Then, without changing the external, macroscopic properties of the gas, there are five possible ways to assign a speed to the molecule in the first vessel, and five for the identical second vessel. When the vessels are joined, however, the number of ways their two molecules can be distributed is not ten, but twenty-five: Each one of the five possible speeds in the first one can coexist with each one of five speeds in the second one. Possibilities and probabilities are multiplicative, while energy and entropy are additive.

This example proves that entropy cannot be identified with simple probability. Boltzmann asked himself: How can a multiplicative property, such as probability, be related to an additive one, such as entropy? Once he posed the question so crisply, the answer, as astonishing as it is simple, came quickly. Every grade-schooler learns its principle without being aware of it: *Whenever you multiply two integers, the numbers of their respective digits add.* 60 *times* 600 equals 36,000—two digits *plus* three digits equals five digits. (The rule sometimes misses by one digit, as in $3 \times 3 = 9$, but that's a negligible error in view of the vastness of the number of molecules in a gas.) So Boltzmann made the bold, inspired guess that entropy equals the number of digits of the corresponding probability. In more formal, mathematical parlance, he expressed this conjecture by equating entropy with the logarithm of probability.

Entropy equals the number of digits of the probability—what a strange mixture of scientific and mathematical ingredients! When Boltzmann first glimpsed it, it must have bewildered him. That this peculiar hypothesis nevertheless turned out to be true is testament to the uncanny, if not unreasonable, effectiveness of mathematics in mirroring nature.

The trick of measuring unwieldy numbers by counting their digits, or taking their logarithms, is not unique to thermodynamics but can be found in other sciences as well. The classification of earthquakes, for example, follows a similar pattern. Instead of describing the strengths of earthquakes directly in terms of seismometer measurements, which differ by enormous factors, the Richter scale counts digits instead. The magnitudes of quakes rated 3 and 4 on the Richter are respectively a hundred and a thousand times greater than that of a small, standard quake rated 1. In this way the Richter scale compresses an interval of numbers ranging from one to a billion down to the much more comprehensible scale from zero to nine.

An even better illustration of this principle comes from physiology, where the scale is not arbitrary, like the Richter scale, but actually meaningful for our senses. For sounds ranging from the barely audible to the threshold of pain, the physical intensity, in terms of energy transported through the air, varies by a factor of a trillion. Our ears and brains cannot deal with such an immense range, so they turn unimaginable multiplicative factors into a more manageable additive scale. In a rough manner of speaking, our ears simply count the digits of the physical intensity ratios and record them as perceived-loudness ratios. Thus a normal conversation may appear three times as loud as a whisper, whereas its measured intensity is actually a thousand times greater. The familiar decibel scale relates perceived loudness to the abstract notion of intensity in exactly the same way that the physical concept of entropy measures the mathematical value of probability.

Once Boltzmann had arrived at his fundamental insight, he proceeded to refine its meaning, and to compare its logical implications with specific, testable experimental evidence. After leaping to his novel theory by induction, he patiently scrambled back down the deductive path to the laboratory. His term "probability" has been called by many names in the subsequent literature (multiplicity, number of microstates, number of outcomes, and number of ways are among the most popular) but the idea is always the same. Probability is a measure of the variety of ways in which the molecules in a system can be rearranged without changing the external, macroscopic properties of the entire system. When the number of ways is high, the arrangement is typical and common—when it is low, the arrangement is exceptional and rare. Among ten coins, a throw of five heads is common, because it can be achieved in 252 ways—ten heads is rare, because there is only one way to do it.

Entropy measures probability, and probability, in turn, connotes disorder. By this reasoning, Boltzmann arrived at the momentous conclusion that *entropy is disorder.* The fact that it constantly increases reflects nature's preference for disorder over order.

My teenaged daughter Madelynn's room illustrates the connection. Order is defined as the state in which all her books rest on their shelves in the same upright position. Madelynn prefers them scattered over the floor in arbitrary orientations; because there are many more ways to achieve such an arrangement than there are ways to put the books on their shelves, the helter-skelter state is more probable than the "orderly" one. In other words, if you were to toss the books into her room at random, they would be far more likely to land on the floor than on the shelves. No matter how she herself may define it, physicists call the more probable state more disorderly. In that way the subjective terms "order" and "disorder" are quantified by association with probability, and identified, respectively, with low and high entropy.

The second law of thermodynamics, applied to Madelynn's room, simply states that in the natural course of events the room has a tendency to become more disordered. To be sure, order can be restored, and entropy lowered, but only at the expense of energy—usually on the part of her parents. In the same way, entropy can decrease, and heat flow uphill, in violation of the second law, but only at the expense of energy: the power cord of a refrigerator proves it.

Another illustration of entropy is amenable to exact numerical calculation. A deck of cards is called ordered when the point values and suits follow each other in a strict methodical sequence, and disordered otherwise. Since this ordered state is unique, it is assigned a very low probability—hence a low entropy. On the other hand, because there is an astronomical number of ways to arrange the cards in a disorderly way, the disordered state of the deck has a very large probability—hence a large entropy. Shuffling an ordered deck will almost certainly take it to a disordered state, but the opposite procedure, randomly shuffling a disordered deck to achieve order, is astronomically improbable. Entropy tends to increase.

With such examples, and others that became increasingly more physical and more complex, Boltzmann checked out the validity of his new accounting system. He calculated how probability increases when heat is added to a body and found that his equation tallied with Clausius's computation of the corresponding increase in entropy; he showed that heat flows downhill because as energy is diluted to a lower temperature, it spreads out over a greater variety of molecular states of motion; and he demonstrated that when a tightly confined gas is allowed to leak into a previously empty vessel, the variety of arrangements it can assume increases so dramatically that the probability of its reversion to the original state is as unlikely as ordering a deck of cards by random shuffling. (Madelynn's room is similar: If all her things were shoved pell-mell into a closet, the room would temporarily assume a high degree of orderliness—a fact she has unfortunately discovered on her own, even though she has never heard of entropy.)

Step by step Boltzmann followed the increase in disorder in the world and recovered the second law of thermodynamics: In reversible processes, like the generation of electricity by a hydroelectric generator, entropy is conserved, but in irreversible ones, like the flow of heat from a cup of tea into the air, the total entropy of the universe increases.

By the same token, the entropy of the world increases irreversibly every time we drive a car. Chemical energy starts out compressed, neat and orderly, into a few gallons of gasoline. Then it is released, turned into heat, and finally converted to work. The process of reconverting the disorderly energy known as heat into the work required to propel the car is costly and inherently inefficient: Two thirds of the energy is wasted, and ends up as disordered, chaotic motion of air molecules in the atmosphere.

After years of painstaking work, Boltzmann achieved the long-desired goal of explaining entropy and the second law. *Qualitatively, entropy is disorder, which has a natural tendency to increase. Experimentally it is measured by the ratio of heat to temperature. Theoretically it is related to the number of digits of the probability.*

With Boltzmann's discovery, the billiard ball model of gas was complete. The three cardinal variables for describing and measuring warmth—heat, temperature, and entropy—acquired purely mechanical interpretations. The language of physics pays tribute to the architects of the model: Heat is measured in joules, absolute temperature in kelvins, and entropy in multiples of a number called Boltzmann's constant.

When Boltzmann traveled to Berkeley, he was at the height of his fame. Students flocked to his lectures, and colleagues throughout the world sought his counsel. His identification of entropy with probability was recognized as a masterpiece of theoretical physics—one of those rare insights he had called "a small opening that allows a hitherto undreamt of outlook on the whole." On his sixtieth birthday during the previous year, a collection of papers had been published in his honor, with contributions from 117 scientists, including some of the era's most illustrious. He had received countless medals and honorary doctorates. But he was not a happy man.

In September of 1906, a year after his California trip, on vacation near the Italian seaside town of Trieste with his wife and daughter, his migraines and depressions overcame him. While the women were off on a swim, Boltzmann tied a short cord to the crossbar of a window in his rented apartment, put a noose around his neck, and hanged himself. His daughter, Elsa, returned to find him dead.

It is sometimes said that Boltzmann was driven to suicide in part by the vociferous opposition of some physicists who misunderstood and misrepresented his belief in the reality of atoms, and much has been made of the irony of the fact that within a year or two Einstein's explanation of Brownian motion would convincingly prove them wrong. But there is little in Boltzmann's writings to bear out this conjecture. His lighthearted description of his journey to California, for example, is not that of a disappointed man. His genius, his fame, and his life force so far exceeded those of the failed suicide Robert Mayer that a comparison between the two men does not seem very instructive. Today, with our enhanced understanding of the sometimes devastating effects of depression, a purely biochemical explanation of Boltzmann's suicide is easier to accept, though it cannot lighten our grief over the suffering of a remarkable human being.

Ludwig Boltzmann is buried in the Central Cemetery of his native Vienna. On his tombstone is engraved, in mathematical shorthand, his immortal testament: "Entropy is the logarithm of probability."

14

APOCALYPSE NOW: THE DISSIPATION OF ENERGY

This is the way the world ends
Not with a bang but a whimper.

—T. S. ELIOT

"Entropy is the logarithm of probability." In this cryptic remark, physicists, who speak the language of mathematics, find beauty and meaning. For them it captures the essence of the second law of thermodynamics, and explains why heat flows only downhill. But for most people, including scientists in less mathematical fields, it is difficult to gain an intuitive grasp of the second law from Carnot's study of the efficiency of steam engines, or Clausius's analysis of the ratio of heat to temperature, or even Boltzmann's astonishing identification of entropy with disorder. While the physicists' conception of entropy was thus proceeding in the direction of the quantitative, the microscopic, and the abstract, another formulation was beginning to capture the imaginations of scientists in other fields and eventually the general public.

Back in 1852, just before Clausius summarized thermodynamics in lapidary fashion, William Thomson, Baron Kelvin, stated the second law in a more comprehensible way. Starting with the observations that the flow of heat from a warm body into a cool one is irreversible, and that even reversible engines discard heat into cool reservoirs, whence it is difficult to retrieve, he concluded that there is "a universal tendency in nature to the dissipation of mechanical energy," and dubbed his hypothesis the "law of dissipation of energy."

Thomson's principle assumes that energy should not only be measured as to quantity, but graded in quality as well—like meat or diamonds. A hot boiler contains high-quality energy, and so does a bowling ball rolling in a well-defined direction with a definite speed. A pot of cool water stores energy of lower quality, as do the bowling pins that skitter in all directions. High quality implies predictability; low quality bespeaks chaos. Quality in energy, in other words, means order, and a tendency to dissipation is a trend toward disorder.

Thomson's law implies that heat flows naturally from hot to cold, and that energy, though always conserved, suffers degradation in all but the most artificial processes. Its most devastating implication is that energy, once degraded, cannot be returned back to high-quality status without the expenditure of more energy—nature's march toward the lowest common denominator is irreversible. The heat that escapes from a cup of tea into the room loses its quality: It cannot be retrieved. The energy compressed into a car's gasoline is transformed into useful high-grade heat in the engine, and partially converted into orderly motion of the car—but a good two thirds of it is frittered away into junk heat at low temperature and tossed out into the atmosphere.

The principle of dissipation of energy was established as a counterpoint to the principle of conservation of energy. Inasmuch as the second law is one of the pillars of physics, this was Thomson's most significant contribution to the science of thermodynamics, and overshadowed his invention of the absolute scale of temperature, his early recognition of the importance of James Joule's work, his tireless search for mechanical models of nature, and even his role as godfather to Maxwell's Demon.

The concept of dissipation resonated with prevailing religious beliefs. Joule had explicitly regarded the first law as evidence of divine order in the universe:

> Nothing is destroyed, nothing is ever lost, but the entire machinery, complicated as it is, works smoothly and harmoniously. And though, as in the awful vision of Ezekiel, "wheel may be in the middle of wheel," and everything may appear complicated in the apparent confusion and intricacy of an almost endless variety of causes, effects, conversions, and arrangements, yet is the most perfect regularity preserved—the whole being governed by the sovereign will of God.

Thomson was deeply affected by this conception, and puzzled how it could be squared with the obvious evidence for the irreversible losses of motive power of Carnot's theory.

In his preliminary draft of the paper on dissipation of energy, he presented his answer in Biblical terms. "The earth shall wax old," he wrote, referring to the 102nd Psalm, which tells of the works of God that shall wax old like garments. If the first law was evidence of the permanence of the Creator, Thomson saw the second one as a reminder of the transience of His creations.

He applied his belief in the separation of the temporal from the eternal to the natural history of the material world: "Within a finite period of time past the earth must have been, and within a finite period of time to come the earth must again be, unfit for the habitation of man as at present constituted."

In a prescient hedge, which would stand him in good stead many years later when he was an old man, he acknowledged that his reasoning was valid only under the laws of nature "to which the known operations going on at present" were subjected.

Hermann von Helmholtz, the Reich Chancellor of German Physics and member of the Berlin Dream Team, soon extended the scope of Thomson's conclusion to the entire universe. All energy will eventually be transformed into heat at a uniform temperature, he argued, and then all natural processes must cease; "the universe from that time forward would be condemned to a state of eternal rest," since there would be no place "lower" down to which heat can flow. Entropy would be at a maximum. Absolute disorder would reign. The primeval chaos of the universe would be reestablished. This melancholy prediction was sufficiently picturesque and disturbing that it even found an audience outside the scientific establishment and entered popular culture in the form of a widespread worry about the "heat death of the universe."

The vague, metaphysical character of the principle of dissipation of energy—the "natural tendency" of energy toward dilution—contrasts curiously with the robust, tangible way in which Thomson described the world. "I am never content until I have constructed a mechanical model of the subject I am studying," he claimed, and, echoing Mayer's insistence on quantification, "I often say that when you can measure what you are speaking about, and express it in numbers, you know something about it; but when you cannot measure it, when you cannot express it in numbers, your knowledge is of a meager and unsatisfactory kind." By this tough-minded criterion, the teleological, quasi-religious nature of his own principle of dissipation must be judged as unsatisfactory indeed.

But vagueness notwithstanding, as long as Thomson stayed within his own discipline of physics, where he was able to survey the multitude of deductions and verifications of a theory, he was on safe ground. It was only when he ventured out into unfamiliar territory and tried to apply the same thematic presuppositions, without the benefit of comprehending the bigger picture for which they form the background, that he overstepped his limitations.

Within seven years of the formulation of the principle of dissipation of energy, an intellectual cataclysm shook the world and plunged Thomson into a controversy he would pursue for the remainder of his long life: On November 24, 1859, Charles Darwin published *The Origin of Species by Means of Natural Selection*. Thomson read it with distaste, and soon found what he thought was its Achilles' heel. As the historian of science Stephen Brush put it trenchantly: "Thomson was a victim of the common delusion that an established scientific theory can be immediately overthrown by citing a single devastating argument against it." The mistake is to imagine that knocking out one single prop under a sturdy trestle bridge will topple the entire structure. Amateur critics of special relativity and quantum mechanics, of which there are legion, suffer from the same misconception, but it is rare to find it in someone of Lord Kelvin's stature.

The point of contention concerned the prodigious length of time required for evolution, which conflicted with Thomson's estimate of the age of the earth. Reasoning within the general doctrine of dissipation of energy, he had followed the flow of heat out of the molten earth into thermal radiation from the surface, compared it with the compensating influx of solar heat, and concluded that the earth had solidified between a hundred and two hundred million years ago. Darwin, on the other hand, spoke of "incomprehensibly vast" stretches of time, and in one instance made a geological estimate of 340 million years. In June of 1861, a year and a half after the appearance of *The Origin of Species*, Thomson wrote disdainfully to a colleague: "I must say that Darwin's principle of estimating 340 million years for some of your ancient South of England headlands seems to me singularly defective; and thousands might be substituted for millions without violating probabilities in connection with any of the facts he mentions." The battle between the biologists and the physicists was engaged.

From the earth Thomson turned to the sun, and endeavored to prove that the fountain of the earth's energy also was too young for evolution to have operated in the way Darwin supposed. He assumed that the sun received heat from two sources: a gradual contraction in size, and the impact of meteors. Both processes are analogous to the heating of a waterfall—potential gravitational energy is converted into heat. The only difference is that in the case of contraction, the material is already present in the body of the sun, whereas meteors carry matter from outer space. On the basis of these assumptions, Thomson concluded that the sun has most probably illuminated the earth for only one hundred million years, and most certainly not for five hundred million years. But again he wisely added the hedge that these conclusions hold "unless sources now unknown to us are prepared in the great storehouse of creation."

So great was the reputation of physicists in general, and of Thomson in particular, that Darwin was beaten into retreat. In frustration he wrote to his rival evolutionist A. R. Wallace: "I should rely much on pre-Silurian times; but then comes Sir W. Thomson like an odious specter." From the third edition of his book, published in 1861, Darwin deleted the controversial estimate of 340 million years, and later he even changed the phrase "incomprehensibly vast" to "how vast." For the rest of the century biologists and geologists, unable to contradict the theoretical onslaught of Thomson and his followers, were forced to suffer what Darwin perceived as the physicists' mathematical arrogance as best they could.

In his presidential address to the British Association in 1871, a year after Maxwell had spoken of descending into the realm of the molecules, Thomson revealed the roots of his passionate opposition to Darwin's theory. It wasn't evolution per se that bothered him, but its implication that the world is ruled by blind chance:

> . . . it did not sufficiently take into account a continually guiding and controlling intelligence . . . the argument of design has been greatly too much lost sight of in recent zoological speculation. . . . But overpoweringly strong proofs of intelligent and benevolent design lie all around us . . . teaching us that all living things depend on one ever-acting Creator and Ruler.

In short, his objection was religious.

The irony of Thomson's position is that although he was the supreme champion of a mechanical interpretation of nature, this conviction extended only as far as physics. When it came to geology and biology, he rejected mechanism in favor of divine intervention—thereby exposing his deepest prejudices in a particularly transparent manner.

Thomson thought that the same power who created the sun, the earth, life, and human beings also directed that they should eventually disintegrate like old garments into primordial dust—but his estimates of the speed with which all these cataclysmic events would unfold were based on insufficient evidence. It was not until after the turn of the century, when Thomson was nearing the end of his life, that he would learn how far he had been from the mark.

The principle of dissipation of energy conformed not only with religious beliefs, but with the Victorian social order as well. The population was sharply divided between the hereditary upper class called "quality" and a vast throng of lower creatures. While the fall from up high through depravity and dissipation was a common theme in literature, the opposite journey, from low to high, was exceedingly rare and required special, artificial effort. (Social climbing was easier in America than in England.) Thomson viewed energy as stratified in the same way as the society in which he lived.

Since degradation is a human and intuitively understandable concern, the principle of dissipation of energy captured the public's attention. The immediate provocation of this interest was the arrival of the end of a century, which frequently triggers a wave of collective apocalyptic hysteria. In the 1890s this frenzy expressed itself in a preoccupation with the heat death of the universe, which seemed to provide a scientific justification for the prevailing mood of fin-de-siècle pessimism and lethargy.

Among the popularizations of this notion was an international best-seller entitled *La Fin du Monde,* published in 1893 by the flamboyant French amateur astronomer Camille Flammarion. Its most lurid illustration is an etching of a dreadful scene on a desolate sheet of ice surrounded by what looks like a wall of towering ocean waves frozen solid. In the foreground a bearded old man in tattered rags, flat on his belly on the ice, is caught in the act of raising his head for a last, hopeless look around. Next to him a younger man with gaunt eyes and Christlike demeanor stands tall and barefoot, desperately clutching the remnants of a threadbare tunic to his body in a futile attempt to ward off the inevitable end, while his scarf flaps in the wind like a broken black wing. The most pathetic member of the little group is the mother with long dark hair who is sitting on a rock behind the men, huddling to protect her son and the baby she clutches in her bare arms. The legend of the whole nightmare is "The wretched human race will perish of cold."

But the end of the world was not quite as close at hand as people thought. Even Thomson—by now Baron Kelvin of Largs—lived to learn how wrong he had been. It happened in 1904 at a lecture at the Royal Institution in London—Count Rumford's emporium of science. Ernest Rutherford, the man who could see atoms in his mind's eye, and who was to discover their nuclei a decade later, was discussing the effect of recently discovered radioactive processes on the temperature of the earth, and hence on its age:

> I came into the room, which was half dark, and presently spotted Lord Kelvin in the audience and realized that I was in trouble at the last part of my speech dealing with the age of the earth, where my views conflicted with him. To my relief, Kelvin fell asleep, but as I came to the important point, I saw the old bird sit up, open an eye and cock a baleful glance at me! Then a sudden inspiration came, and I said Lord Kelvin had limited the age of the earth, *provided no new source was discovered*. That prophetic utterance refers to what we are considering tonight, radium! Behold! The old boy beamed upon me.

Rutherford merely pointed out the first new source of energy of the twentieth century: radioactivity. Later, fusion, fission, and even the evaporation of black holes were to be added to the list. Together they would push the advent of the heat death of the universe so far beyond conceivable geological and biological time spans that scientists and popular writers began to lose interest in its imminence. When you're young, you don't worry too much about death or old age.

To be sure, no objection in principle has been raised against the prediction of heat death. However, the complexities of the history of the universe from the Big Bang to its ultimate fate now seem so overwhelming, and the uncertainties of our predictions so great in view of the impossibility of their experimental verification, that the problem has lost its ominousness.

Rutherford's newly discovered energy ultimately comes from an unexpected and almost inexhaustible source—nothing less than the combined mass of all the stars and galaxies, of all the matter, in the universe. But Einstein's proposal that energy and mass are interconvertible in his famous formula $E = mc^2$, which was published just a year after Rutherford's lecture at the Royal Institution, resulted not only in an indefinite postponement of the heat death, but also in a radical reinterpretation of the concept of energy.

15

$E = MC^2$

The visible world is neither matter nor spirit but the invisible organization of energy.

—HEINZ PAGELS

$E = mc^2$ is the only mathematical formula in *Bartlett's Familiar Quotations.* Its symbols, however many ways they may be pronounced, are common in all languages, so the formula is by a wide margin the most familiar quotation in the entire collection. It is probably also the least understood. Nevertheless, in books and magazines, on stamps, in cartoons, in advertisements, on T-shirts, and even in casual conversation, it appears as a kind of magical incantation, a spell uttered to invoke the whole incomprehensible and often terrifying world of atomic physics and nuclear weaponry. At the same time it is the basis, in its precise, mathematical context, for some of the most impressive technical achievements of physical science.

$E = mc^2$, as the simplest mathematical formula for energy, joins a long list of other expressions for the same basic quantity. As soon as Robert Mayer conceived of the notion of a new conserved quantity, he realized that every one of its innumerable forms must have its own unique equation. If energy is defined as the ability to cause something to move, then a puck sliding across the ice possesses energy—

energy of motion, or kinetic energy—because it could hit another puck and make it move. The derivation of the correct formula for kinetic energy caused Mayer much pain—he got it completely wrong in the paper that was rejected, and was still off by a factor of two in the one that made it into publication. But eventually the muddle was untangled and everyone agreed on the form of the equation—one half the mass times the square of the speed—which children learn in school today.

Another type of energy is stored in a weight at rest, but raised above the ground. This kind, called potential energy, exists by virtue of the fact that the object *could* acquire motion if it were dropped. It was familiar to Mayer as "fall-force" and to Joule as "effort," and both knew how to calculate it as weight times height. To the formulas for kinetic and potential energy, they added the mechanical equivalent of heat to account for the fact that heat is motion, and through the mediation of a steam engine, for example, can cause motion. Furthermore, Joule had formulated an equation for the energy consumed by an electric motor. After these preparatory steps, physicists scurried to find ways of calculating the energy contents of stretched springs, orbiting planets, electrically charged conductors, electromagnets, water waves, light, sound, batteries, body chemistry, and every other system that moves or has the potential to cause motion. In 1905, when the aging Ludwig Boltzmann was explaining entropy in California, the twenty-six-year-old patent clerk Albert Einstein ushered in twentieth-century physics by adding the formula $E = mc^2$ to the list.

Actually, the discovery of the equation was not as spectacular as its current fame suggests. In fact, $E = mc^2$ was an afterthought. Although it is a basic consequence of the theory of special relativity, Einstein hadn't yet found the formula when he published his first paper on relativity in June 1905. Three months later he submitted a second article, a mere three pages unadorned by footnotes or references, in which he derived the equation from concepts and calculations in the major work. Einstein realized its potential significance, though privately he hedged a little. "This argument," he wrote to his friend Conrad Habicht, "is amusing and attractive; but I can't tell whether the Lord isn't laughing about it and playing a trick on me."

In print, however, Einstein was characteristically bold and prescient. While lesser scientists might have speculated bravely in private but hidden behind a mask of cautious objectivity in public, Einstein suppressed his qualms and risked his reputation with the astonishing assertion that matter and energy are equivalent. It was also typical of him to end the article with a guess about a way to check the formula experimentally—the only reliable way to establish its correctness. He suggested that comparing the loss of mass of radioactive substances with the energy they released would show that this process obeys the dictates of $E = mc^2$. Thirty years later, when the means for measuring radioactive decay had caught up with Einstein's hunch, it became possible to prove that the formula tallies exactly with the evidence. But by then physicists had long been persuaded of the equation's truth.

The meaning of $E = mc^2$, which relates energy, mass, and the speed of light in a neat little package, can be grasped without understanding its derivation. Mass, denoted by m, is a measure of inertia—the tendency of things to resist being set in motion when they are at rest and to resist changes in velocity when they are moving. Inertia is a symptom of nature's laziness, and mass allows us to measure that trait. To the physicist, inertia and mass are expressions of the same property, one qualitative and the other quantitative. Indeed, Einstein chose to use the qualitative word in the title of his paper "Does the Inertia of a Body Depend Upon Its Energy Content?" and not to mention mass at all.

The surprising element that unites mass with energy and makes the formula $E = mc^2$ so strange and powerful is the speed of light, *c*. In the classical mechanics perfected by Isaac Newton, motion and mass have nothing to do with light or its speed. The behavior of light was understood in terms of Maxwell's theory of electromagnetism, which held that light consists of oscillating electromagnetic fields. It was Einstein's achievement to discover that the two theories are in fact inconsistent. A bullet shot from the nose of a fast jet adds the speed of the airplane to its own, whereas a beam of light shining from the searchlight of the jet is clocked at the same speed as light shining from a stationary source. (This strange, counterintuitive fact is a consequence of the laws of electromagnetism and has been experimentally verified. It cannot be explained, but must be accepted.) In order to reconcile the description of bullets with that of light, Einstein combined the two pertinent equations for calculating speed into one. By way of this back door, the constant *c* (approximately 186,000 miles per second) sneaked into the description of all motion. And, since energy is a measure of motion—or of the possibility of motion—it too must be described with reference to *c*.

Einstein realized that the factor c^2, which is almost unimaginably huge by human standards, renders the energy contained in a pound of butter, or a pound of anything else, immense beyond belief. But he couldn't put that knowledge to use. Science, in the winter of 1905, was like a world containing both water and ice with no means of freezing one or melting the other. Einstein showed that energy and mass are different forms of the same fundamental entity, but neither he nor anyone else knew how to turn one into the other.

Two hundred years earlier, Newton had suspected that such transformations occur. Writing about light, which is a form of pure, matterless energy, he observed, in query 30 of Book III of his *Opticks*, "The changing of bodies into light, and light into bodies, is very conformable to the course of Nature, which seems delighted with transmutations." In due course, it was found that nature not only delights but positively wallows in transmutations of matter into energy and energy into matter, all obeying the law $E = mc^2$. Cosmology tells the story of the freezing of the Big Bang's primordial energy into the most primitive particles, called quarks and leptons, which eventually congealed into the atoms and molecules of our everyday world. Astrophysics records the subsequent melting of matter in the sun's interior back into energy that reaches us in the form of light and animates our world. Even geology is not exempt, as Ernest Rutherford had explained to the aged Lord Kelvin. In the earth's crust the transmutation of matter into energy is silently continuing as the decay of radioactive elements liberates heat and warms us imperceptibly. All these metamorphoses add up to a spectacular fulfillment of Newton's vision.

Eventually technology learned to emulate nature: Mass becomes energy in nuclear reactors and atomic bombs, where the natural processes of fusion and fission convert a tiny percentage of the mass of the fuel into a blaze of heat and radiation. Just as challenging intellectually, though not of much practical use, is the reverse process, the conversion of E into m. Colliding beam accelerators, for example, convert the kinetic energy of fast light projectiles racing around in their tunnels into the masses of the heavy new particles that are born in the explosive collisions of the beams.

But Einstein's formula also has a deeper philosophical implication. At the most fundamental level it implies that energy *is* mass, and vice versa. From that perspective, the square of the velocity of light is a mere conversion factor for changing joules to kilograms, similar to the factor 100, which performs the trivial transformation of dollars into cents. Since inertia seems to be the very opposite of energy, with its connotations of motion, vivacity, and change, their equivalence—like that of ice and water—is all the more profound for its paradoxical quality.

With the rebirth of energy as mass, the law of its conservation returned to its origin. After Robert Mayer had come home to Heilbronn and clambered about three quarters of the way up the inductive path to his historic discovery, he remembered Lavoisier's law of conservation of matter, which guarantees that the total mass in a chemical reaction is preserved. (We now realize that this can't be quite right if some heat is evolved, or some of the energy is stored away in chemical form, but on account of the enormity of the factor c^2, which represents the price for changing mass into energy, the effects of such processes on the masses of atoms are too small to be measured.) Mayer saw the relationship between the conservation of energy and the conservation of mass as an analogy—Einstein elevated it to an identity. How damaging this discovery would have been to Mayer's metaphysical belief in the superiority of spirit over matter is impossible to guess, but for us it has the effect of unifying two great laws of nature into one.

The law of conservation of energy, reborn as the law of conservation of mass/energy, has established itself as one of the few unshakable theoretical guideposts in the wilderness of the world of our sense experiences. In scope and generality it surpasses Newton's laws of motion, Maxwell's equations for electricity and magnetism, and even Einstein's potent little $E = mc^2$. It survived not only the storms of the quantum revolution, which transformed the crisp granularity of the atomic realm into a miasma of probability, but also the flood of cosmological discoveries that shattered ancient preconceptions about the permanence and simplicity of the universe. At all scales, from the unimaginably small to the inconceivably large, the law holds sway. It comes as close to an absolute truth as our uncertain age will permit.

The second law of thermodynamics, on the other hand, continued to trouble Einstein. Before 1905 he had failed three times to derive it from mechanics and probability. After the advent of his theory of special relativity, he turned his full attention to his theory of gravity, which would see the light as general relativity in 1916, and stopped research on thermodynamics. But the irreversibility implied by the second law could not help but exert its mysterious appeal on someone as deeply committed to the search for the meaning of time as Einstein. After a long, productive, and turbulent career, a half century after his miracle year of 1905, he revealed that near the end of his life he was as mystified by the flow of time as he had been in his youth.

16

HERE AND NOW: THE ENIGMA OF TIME

The essence of nowness runs like fire along the fuse of time.

—GEORGE SANTAYANA

On the fifteenth of March 1955, Albert Einstein wrote a note of condolence to the family of his friend Michele Besso, who had died a week earlier. The letter begins with an excruciating confession about his own life: "What I admired most about Michele was the fact that he was able to live so many years with one woman, not only in peace but in constant unity, something I have lamentably failed at twice." Further on, after recalling his meeting with Michele almost six decades earlier when they were university students in Zurich, and the difficulty of keeping in touch through the intervening years, Einstein rouses himself from his melancholy and soars to a higher plane. His concluding passage rings with pure spiritual power: "So in quitting this strange world he has once again preceded me by a little. That doesn't mean anything. For those of us who believe in physics, this separation between past, present, and future is only an illusion, albeit a stubborn one." Less than four weeks later Einstein followed his lifelong friend out of this strange world.

His characterization of time as illusory was not a dying man's lapse into mysticism, but a carefully considered scientific conviction about the structure of the world. The enigma of time held an enduring fascination for Einstein—as it does for everyone. Why is there a difference, apparent to every child, between the past and the future? Does time always flow in the same direction, like a river, or can its course be reversed? And, most mysterious of all, what is this experience we call Now, that cusp between the upswelling wave of the future and its ebbing away into the past that lies at the very core of consciousness, yet escapes every attempt to capture it?

Since Greek antiquity, when philosophers began to turn their attention to the study of motion—the change of an objects's position with the passage of time—space and time have formed the scaffolding that supports the laws of physics. As Galileo and Newton, and later Einstein himself, delved ever deeper into the analysis of motion, surprising and increasingly compelling analogies began to link those two disparate concepts until, in the end, they fused together. In 1908, the mathematician Hermann Minkowski was able to announce to an audience of physicists and physicians meeting in Cologne: "The views of space and time which I wish to lay before you have sprung from the soil of experimental physics, and therein lies their strength. They are radical. Henceforth space by itself and time by itself are doomed to fade into mere shadows, and only a kind of union of the two will preserve an independent reality." Minkowski's abstract conception of space-time is what Einstein had in mind when he called the separation of past and future an illusion.

From the perspective of relativity, past and future have no more fundamental significance than left and right, or forward and back. In that context, moving back and forth in time is no more mysterious than walking around in a room. As your spatial point of reference, your Here, you may pick the corner of your desk, the North Pole, or the center of the sun. If you leave that point, even in your imagination, you can always return to it from wherever your motion has taken you. For Minkowski, time was simply a fourth dimension, similar in nature to the three we live in. Every particle in the universe is described by a long, thin thread, called its world-line, that snakes its way through four-dimensional space-time and represents the particle's entire history. Each point on the world-line is characterized by a specific location in space, and the instant at which the particle occupies that location. The totality of particles in the universe is represented by a thick tangle of intertwined world-lines that fills space-time like a vast ball of spaghetti. In this picture, the future is as manifest and as accessible as the past. Forward and backward in time are matters of convention, like forward and backward in space. As a reference point, Now is as arbitrary as Here.

But of course that's not how the world appears to us. We always move forward in time; later follows earlier as inexorably as thunder follows lightning. As for a point of reference—we can point to a spot and call it Here, and invite our friends to come and examine it, but when we attempt the equivalent experiment of fixing a Now, it vanishes as swiftly and irretrievably as a puff of smoke in the wind.

I tried to capture a moment once, when I was a boy growing up in Switzerland; I must have been about twelve. I was returning home by train one evening from a visit to my grandparents in Basel, and began to mull over the elusiveness of time. I resolved to stop its passage by fixing one moment, one isolated, specific Now that would forever remain frozen in my memory. Since I had taken the same trip many times before, I knew that we would soon pass a castle on the right, a romantic little chateau with turrets and archways made of yellow bricks outlined in red and set back in a clearing in the woods. That castle, called the "Feldschlösschen," or little hunting lodge, appealed to my sense of adventure, and figured in all sorts of imaginary exploits. Much later I found out that it actually belongs to a famous brewery, and serves as its trademark.

Straining my powers of concentration to their limit, I prepared to fix that moment. At the crucial instant, just as the train sped past the clearing, I murmured "*Now,*" and skewered the moment like a lepidopterist's specimen. During that fleeting moment I was totally conscious of myself, my thoughts, my feelings, and my surroundings—the familiar compartment of the Swiss train, the rhythmical drumming of the wheels, the warm wind rushing by the open window, the beech forest with its clearing, and the chateau itself illuminated by the slanting rays of the setting sun. And then, in a flash, the moment of clarity vanished into the vast, indifferent sand dunes of the past.

I will never lose that moment as long as I live. By itself, it was quite unexceptional, but it was made singular by a curious boy who happened to choose it as his point of reference in time. Later, to complete the space-time description of the event, I looked up the location of the Feldschlösschen and found that it is situated just across a hill from the town of Aarau, where Einstein attended high school and experienced the first vague intimations of what would later blossom into the theory of relativity.

Einstein brooded about the transitoriness of time. He once told a philosopher of his acquaintance that the problem of the Now worried him seriously, and that he was painfully disappointed it could not be grasped objectively by science. He felt that there is something essential about the Now that is just outside the reach of science. When we understand why Now is different from all other moments, he hoped, the distinction between past and future will finally reveal itself as a stubborn human illusion.

As it happened, the first solid clue to the true nature of time came not from poetry or philosophy or any of the humanities, nor even from psychology, but from the most applied frontier of physics, where it bumps up against technology and engineering practice. It was the goal of improving the steam engine that led Sadi Carnot, the founder of thermodynamics, to invent the abstract notion of a reversible engine—one that can run backward as easily as forward. Generalizing this conception, he divided all processes into "reversible" and "irreversible," and thereby created a language for the scientific analysis of the enigma of time.

In cinematographic terms, a reversible process is simply one in which a movie can be run forward or backward with equal plausibility. Imagine, for example, a sequence of collisions in a perfect billiard game, in which neither rolling nor bouncing entail any loss of energy. Such a game is realistically unattainable, but it can be approximated, and Carnot had no difficulty in imagining it. If a movie of this sequence is projected onto a screen, there is no way of telling whether the projector is running forward or backward—the entire process is reversible. Or consider the flight of a baseball through a vacuum. As long as you see neither the hit nor the catch, no measurement, no matter how clever, can decide whether the film is running forward or backward.

A film of a windowpane being smashed, on the other hand, or of a drop of ink dispersing in a glass of water communicates its correct direction immediately. Richard Feynman's test for irreversibility was that when you run these films backward, the audience begins to laugh. In this fashion, everything that happens in nature can be labeled reversible or irreversible. With a little bit of thought, one can even devise ways of measuring the degree of reversibility or irreversibility. Thus the fall of an object through the air is highly reversible, because the subtle differences between its downward and upward motions caused by air resistance, though measurable, are minute.

Carnot invented the concept of reversibility in an effort to distinguish between avoidable and unavoidable losses of efficiency in steam engines, but its significance far exceeds that limited, technical application. The real payoff for its introduction is the light it sheds on the meaning of time. For most, if not all, of the mystery of the Now derives from the irreversibility of the world.

What was it, precisely, that made my moment at the Feldschlösschen so magical? Most of the external conditions of the event could have been re-created, at least approximately, on the next day. If I had taken the same compartment of the same train, and the weather hadn't changed, I could have counted on Swiss punctuality to pass the same spot at the same time. I would have experienced the same scene, and could have duplicated my Now—thus robbing it of the uniqueness that accounts for much of its mystique.

But of course, closer examination would have revealed countless subtle changes in every part of the little snapshot. I would have been imperceptibly taller and older, my mind would have been enriched by the memories of the previous day, the grass and trees would have been a day further along in their yearly march toward autumn—the very molecules of the wind would have been new, and the surface upon which I sat would have presented a vastly different structure to a microscope capable of distinguishing atomic details. All these differences would have been evidence of irreversibility. No, the Now of my childhood is sacrosanct, unique, and gone forever.

For Einstein, the eternal laws that guide billiard balls, electrons, and planets along reversible paths provided a more secure foundation for understanding the world than our fickle perceptions. He realized sadly that these laws fail to capture the human experience of the Now. The phenomenon of irreversibility, on the other hand, which is responsible for most of the Now's mystery, is eminently suitable for scientific inquiry. The irreversibility of processes such as the breaking of a pane of glass and the flow of heat from a hot object into a cooler one imposes on time a sense of direction that is absent from the three dimensions of space. That unique forward direction of the flow of time, time's arrow, is the necessary prerequisite for the division of time into past, present, and future. It is the source of our stubborn illusion of progress. And, as Einstein well knew, it receives its most cogent scientific expression in the second law of thermodynamics.

17

TIME'S ARROW

Time flows on, never comes back. When the physicist is confronted with this fact he is greatly disturbed.

—LEON BRILLOUIN

You can't get an odd number by adding even integers—no matter how many of them you string together. Nor, it would seem, can you generate an irreversible arrow of time by compounding reversible processes. And yet, that's what nature seems to accomplish when it entangles us in the irreversible unfolding of events, which in turn consist of countless atomic processes that are perfectly reversible. This fundamental paradox cloaks the second law of thermodynamics in a shroud of mystery and ambiguity, while the energy conservation of the first law shines forth as a steady beacon of certainty in an uncertain universe.

Newton's and Einstein's laws of mechanics, Maxwell's equations of electromagnetism, the quantum theory of atoms, and even, with one rather insignificant exception, the modern descriptions of elementary particles in terms of quarks are time-symmetric. A film of any elementary process that adheres strictly to these laws and is projected backward depicts a perfectly normal, unremarkable, physically plausible event. How does the obvious irreversibility of the world emerge from its reversible building blocks?

This question has dogged physicists ever since Ludwig Boltzmann discovered the meaning of entropy in the atomic constitution of matter. Actually, Boltzmann himself suggested an answer, and most scientists believe it, but a rigorous mathematical proof of the second law, starting with a reversible description of individual atomic processes, has not yet been achieved, in spite of the determined attack by an army of theoreticians that counted the young Einstein as an early recruit. Whether such a proof will ever be found, or whether, on the contrary, the second law will retain the independent status it now enjoys alongside the other fundamental axioms of physics, is not known.

Boltzmann's explanation of the origin of irreversibility, like his identification of entropy, rests on the calculus of probabilities, which is to say on chance and large numbers. How randomness can impose upon large numbers of reversible events the semblance of irreversibility is illustrated most clearly by a game invented in 1907 by the physicist Paul Ehrenfest, who had studied with Boltzmann and adopted his way of thinking. Let us call on Maxwell's Demon to lend us a hand playing it.

Imagine two large urns and a hundred Ping-Pong balls numbered from one to a hundred, like the ones used to draw lottery numbers on TV. We will also need a roulette wheel with a hundred slots, or some other means of generating random integers from one to a hundred. The rules of the game are simple. To begin with, the Demon drops seventy-five randomly chosen balls into one urn, and twenty-five into the other. He then picks a number at random, fishes out the corresponding ball from whichever urn he finds it in, and plops it into the other urn. Then he repeats the procedure.

Suppose that the first number he picks is 31. Since the first urn is fuller, chances are three to one that ball number 31 is in it. Thus the likelihood that the first transfer tends to empty the fuller urn and fill the emptier one is three times larger than the converse. Depending on what happens in the first turn, the probabilities for the next ball—say number 41—are a tiny bit more or less than three to one. Therefore the first urn will gradually begin to empty and the second one to fill up. After many draws and transfers the indefatigable Demon will inform us that the urns are both approximately half full, and that except for minor statistical fluctuations they seem to remain so. Since a ball is always more likely to be in the urn that is more than half full, the trend will always continue to be toward a 50-50 ratio.

Ehrenfest's model accurately reflects many features of the real world. The number of constituents is large, the number of chance interactions (i.e., transfers between urns) unlimited. Each individual transfer is perfectly reversible, since there is nothing in the physical arrangement or in the rules to favor one urn over the other. Yet out of this symmetrical setup an arrow of time emerges as if by magic: The trend toward equal occupation of the urns defines its direction.

The double-urn game is a convincing metaphor for the flow of heat. If two identical metal blocks, one warmer than the other, are brought into contact, heat will flow downhill irreversibly until, except for statistical fluctuations, they reach equilibrium at the same temperature. The advantage of Ehrenfest's game is that its progress can be predicted by means of the calculus of probabilities. A numerical simulation by computer is even easier, and yields graphs of occupation numbers that start at 75-25, jiggle up and down erratically, and on average approach each other to meet at equilibrium.

But a dreadful skeleton lurks in the closet. Imagine that the Demon keeps on playing day after day, week after week, year after dreary year. Is it not possible in all that time that a series of fluke draws will produce a run—a gigantic statistical fluctuation—resulting in seventy-five balls in one urn and twenty-five in the other? In fact, it is not only possible, but *dead certain*. In an infinite series of numbers, all possible sequences will occur not just once but infinitely often. From time to time, each urn will hold seventy-five balls, or even one hundred, and occasionally ball number 37 will be transferred back and forth three hundred times in a row. Everything that is possible is certain to occur—eventually. That's the wonder of infinity.

And that is also the paradox of time's arrow. No sooner have we found a persuasive mechanical model of the irreversible flow of heat than it turns around and bites us. We are forced to admit that occasionally heat can flow uphill. A cup of tea, sitting out on the kitchen table to cool, and left to itself long enough, will one day spontaneously begin to boil. The ink dispersed in a glass of water will reassemble into a single drop. Time itself will seem to flow backward. We must conclude, in short, that the second law of thermodynamics is not absolute, like the first law, but only statistical in nature.

It bothered Maxwell, that little qualification "only." Without really understanding Clausius's definition of entropy, or its interpretation by Boltzmann, Maxwell wrote to his friend Peter Tait: "Concerning Demons. 1. Who gave them their name? Thomson. 2. What were they by nature? Very small BUT lively beings incapable of doing work but able to open and shut valves which move without friction or inertia. 3. What was their chief end? To show that the 2nd Law of Thermodynamics has only statistical certainty." Like all great physicists, Maxwell smelled the truth long before it was possible to prove it, and dispatched his secret agent to verify it.

To return to Ehrenfest's game, its most important feature, and perhaps one of the most difficult to grasp, is the immensity of the numbers involved. The Demon would have to wait for an awfully long time before he could reasonably expect to encounter a refilling of one of the urns, so the probability of witnessing such a fluctuation in our lifetime is correspondingly minute. But instead of pursuing this rather artificial calculation, let us turn to a more realistic example.

Consider a quart bottle full of air molecules at room temperature. Instead of a hundred, they number almost a trillion trillion—that's one followed by twenty-four zeros. Quantum mechanics allows us to count the number of ways in which the positions and speeds of these molecules can be rearranged without changing the volume, pressure, or temperature of the bottle. (In Ehrenfest's model this is the number of ways of distributing a hundred balls into two urns.)

The mathematical physicist David Ruelle, who thinks deeply about chaos, statistics, and the properties of large numbers, and in addition writes with exquisite grace about these difficult subjects, has estimated this number, which we shall call *M* for "monstrous." *M* far exceeds the number of seconds that have elapsed since the Big Bang, the number of atoms in the universe, and the diameter of the visible cosmos expressed in inches. In fact, Ruelle reports that *M* is so far from ordinary intuition that it elicits intense revulsion in some people and inordinate enthusiasm in others. He was not able to write it out in full because that task would have consumed more than his lifetime. Instead, he resorted to counting its digits.

According to Ruelle, if you are satisfied with merely estimating the number of *digits* of *M*, you get another number, essentially the entropy of a quart of air, that has twenty-two *digits* itself. Mathematicians would say that *M* is approximately equal to ten to the power ten to the power twenty-two.

Now imagine that by normal collisions the air molecules rearrange themselves at random among these *M* configurations. Pick a duration for a typical rearrangement—a second or a trillionth of a second, it doesn't much matter. If the molecules cycle at random through all their configurations, how long would it take to reach some particularly unusual state—with all of them in a blob on the bottom of the bottle, or some other fluctuation far from equilibrium?

The answer is unimaginable because *M* is unimaginable. It exceeds the longest measure of time we can grasp, the age of the universe, by so much that analogies fail. The time is so long, in fact, that Boltzmann called it infinite. Accordingly he suggested that for practical purposes, the trend is *always* toward equilibrium, never away from it—that extremely rare fluctuations *never* occur. In dealing with outrageously large and unimaginably small numbers, mathematicians and physicists part company.

But even though the numbers support Boltzmann's statistical interpretation of the arrow of time, questions of principle remain. In particular, it may be that the double-urn model leaves out important features of the real world that change the conclusions. For example, each transfer of balls is completely independent of the preceding one. But in a real gas this is not true. Molecules exert forces on each other, and these spoil the independence of different arrangements. Consider two molecules that were far apart and are brought together. If they happen to attract each other, the probability that they will move apart again in some future rearrangement is diminished. Correlations among actual events invalidate statistical computations based on the assumption of independence.

When correlations are brought to bear on the calculation, together with the demands of quantum mechanics, which smears Newton's crisp particle trajectories into clouds of probability, and special relativity, which imposes minimum time delays on all interactions, not to speak of such practical problems as the impossibility of absolutely isolating a quart of air from its environment, and of observing its molecular configurations with sufficient precision to make predictions—when all these complications are taken into consideration, the similarity between Ehrenfest's model and Ruelle's bottle of gas begins to evaporate. Some real effects in the gas actually strengthen Boltzmann's hypothesis, others weaken it, but together the challenge of mastering them assures a long, eventful future for theoretical physics. Will the second law become a mere mathematical theorem, derived from more fundamental axioms, or will it retain its independent status as what the astrophysicist Arthur Eddington called "the supreme Law of Nature"? Nobody knows.

For all its imperfections, the double-urn game draws attention to one aspect of the world that is of singular significance. In order for an arrow of time to develop, the game must start with an exceptional distribution of balls such as 75-25. Had the Demon started instead near the equilibrium distribution of 50-50, no forward direction in time would have become noticeable.

A distribution such as 75-25 is called more orderly than 50-50 because there are fewer ways to realize it. This can be seen most clearly in the extreme case of 100-0, which can occur in only one way. By way of contrast, the number of ways of dividing one hundred numbered balls into two equal groups has about thirty digits. In thermodynamic terms, the more orderly arrangement corresponds to lower entropy, so the requirement for Ehrenfest's game is that it must begin with a state of very low entropy. Only then can you expect the entropy to grow in accordance with the second law.

Applied to a real, irreversible process, such as the cooling of a cup of tea, the demand for an exceptional initial condition is self-evident. Without it, nothing happens. A movie of a cup that begins at room temperature—in equilibrium—is like that legendary Andy Warhol film of a man (who happened to be my college chum John Giorno) sleeping for five hours: Even John wouldn't be able to tell whether the projector is running backward or forward. In order to demonstrate the arrow of time, a cup of tea must obviously start out hotter (or cooler) than the surrounding air.

But in one special case the reasonable insistence upon a simple, orderly initial condition is not so obvious. It has, in fact, mind-boggling implications. That case is the universe, to which Rudolf Clausius first boldly applied the laws of thermodynamics. If Boltzmann's idea is correct, the universe must have started in an unbelievably orderly arrangement, from which it has degenerated into disorder ever since. In the nineteenth century, before the discovery of the Big Bang, and before cosmology became a respectable branch of physics, Boltzmann was reduced to mere speculation about the meaning of this extraordinary conjecture, but in his characteristically fearless fashion he waded right in. "That the world started . . . from a very improbable initial condition, this can be counted among the fundamental hypotheses of the whole theory," he wrote, "and it can be said that the reason for this is just as little known as the reason why the world in general is precisely so and not otherwise."

Today, a hundred years later, we are a bit closer to understanding the meaning of the initial condition. In the beginning, when all the universe's energy was compressed into a tiny ball smaller than a marble, order was at maximum—just as it is in Madelynn's room when everything is shoved into her closet. Furthermore, quantum theory has taught us to pay attention to the other end of the arrow, the modern equivalent of heat death, as well. Murray Gell-Mann, the American Nobel laureate and architect of the quark theory of matter, complemented Boltzmann's thinking when he wrote recently: "The thermodynamic arrow of time can be traced back to the simple initial condition of the universe and the final condition of indifference in the quantum-mechanical formula for probabilities of . . . histories of the universe." Of course he also explained that we are a long way from being able to write down that formula.

The extreme orderliness of the initial condition required by the arrow of time is a boon for the fledgling science of quantum cosmology. In order to figure out how the universe evolves, we have to know how it started. Since we cannot go back to observe its initial condition directly, we are grateful for a strong theoretical argument that tells us how things started out. As we await the gradual development of quantum cosmology, we recognize in Boltzmann's idea a triumph of human reasoning. It uses the second law of thermodynamics as a magic carpet to whisk us from our own commonplace experience of the world, backward to a time ten or twenty billion years ago, to reveal a startlingly tranquil vista of the infant universe. Few scientific arguments can match its scope and daring.

18

FOUR OBITUARIES FOR THE DEMON

Let us stop here and be grateful for the good old Maxwellian Demon, even if he does not assist in providing power for a submarine. Perhaps he did something much more useful in helping us to understand the ways of nature and our ways of looking at it.

—W. EHRENBERG

The second law of thermodynamics imposes its direction on cosmic history: It prevents the universe from reversing its course. No wonder, then, that scientists fear the Demon, that irrepressible goblin who seems to be able to violate the second law and buck the flow of time. For physicists, the only good Demon is a dead Demon. But he refuses to stay that way, as the recurrent reports of his death demonstrate.

1914: The Mechanical Demon (Age Forty-seven) Succumbs to Heat

The Polish physicist Marian von Smoluchowski noticed that an automated Demon heats up so much that a random, uncontrollable tremor, called its Brownian motion, prevents it from carrying out its assigned task of violating the second law.

But even in death, the mechanical Demon managed to enlighten and instruct us. His demise on account of

Brownian motion drove home the unexpectedly important role of statistical fluctuations, or deviations from average behavior. Without such rare but inevitable excursions—which troubled my daughters when their ten pennies failed to come up with six heads—motes of dust would not dance in a ray of sunlight, nor would grains of pollen skitter about in a drop of water. Without statistical fluctuations, a large intruder in a crowd of small molecules would feel the same average pressure from all directions at all times, and would remain in equilibrium. Brownian motion, in other words, would cease. If that were the case, Maxwell's mechanical Demon, the pressure valve, could be rigged to extract work from a double vessel full of gas without discarding waste heat, and the second law would fail.

The Demon's death thus carried the subtle and unanticipated message that departures from the average, far from being unwelcome side effects that are best ignored, are actually *required* to protect the validity of the law of entropy. No discussion of the second law of thermodynamics could henceforth ignore consideration of the effect of fluctuations—thanks to the defenestration of the mechanical Demon.

In the end Smoluchowski hedged his bet. After imposing the death sentence on the mechanical Demon, he revived Maxwell's original creature by allowing that intelligence might somehow circumvent the effect of Brownian motion. Thus the *intelligent* Demon popped back into the house of science.

1929: Measurements Fell the Intelligent Demon (Age Sixty-two)

In 1925 the brilliant Hungarian physicist Leo Szilard earned his Ph.D. from the University of Berlin with a dissertation on the foundations of thermodynamics. In later years he was destined to become one of the seminal thinkers of his generation—a quiet genius whose name is not well known to the public, but whose ideas inspired his more famous colleagues. With Albert Einstein he took out a Swiss patent for a novel refrigerator. With the Nobel laureate Enrico Fermi he held the patent on the first nuclear reactor—albeit without reaping any significant monetary rewards. With his fellow Hungarian Edward Teller he foresaw the possibility of atomic weapons, and drafted the letter which, over Einstein's signature, persuaded President Roosevelt to initiate their actual development. And after World War II, Szilard became one of the founders of the influential international Pugwash Conference on nuclear disarmament, which went on to win the 1995 Nobel Peace Prize.

All that lay in the future when Szilard undertook to kill the Demon in a 1929 paper entitled "On the Decrease in Entropy . . . by the Intervention of Intelligent Beings." The entropy decrease in the title referred to an increase in orderliness, and thus a violation of the second law. Since Szilard believed in the second law, he decided to submit the Demon's actions to a scrupulous examination.

To this end he gave up on Maxwell's homunculus, and instead invented various imaginary contraptions that are almost purely mechanical, but require the intervention of an intelligent being at some well-defined point. (These devices were so cleverly conceived that they continue to provoke debate to this day.) Szilard discovered that intelligence is invariably invoked to make some *measurement*. Maxwell's own sorting Demon, for example, measures the identity, position, speed, and direction of travel of a molecule before he decides whether to admit or block it. In this act of measurement, Szilard found the safeguard of the second law.

"Measurements themselves are necessarily accompanied

by a production of entropy," he concluded, because he assumed (incorrectly, as it turned out much later) that every real measurement is accompanied by some waste. And when Szilard examined the measurements in his examples, he found that in every case the intelligent being dissipates precisely the amount of heat required by the second law. The Demon seemed dead again.

That measurements should waste energy seems intuitively obvious. Activities like counting pennies, weighing potatoes, and surveying land take work, and cause sweat. The mechanical aspects of such activities are riddled with sources of friction, resistance, and waste. Furthermore, even the act of remembering the measured numbers is accompanied by physical changes in a living or mechanical brain, and must be accounted for. Szilard's argument seemed cogent.

Unfortunately, no matter how many ingenious examples he thought up, they did not constitute a proof. For a mathematical proof of the second law along Szilard's line, you would need a general, quantifiable, and universally applicable definition of what is meant by the vague word "measurement," and that doesn't exist. Szilard's analysis was suggestive, but stopped short of killing the Demon.

In spite of its shortcomings, Szilard's work remains an important contribution to our understanding of the meaning of physical theories. In the same year that Szilard obtained his doctorate, young Werner Heisenberg was inventing the field of quantum mechanics, and set in motion a chain of reasoning that would focus on the act of measurement as a fundamental process of crucial significance for the formulation of theoretical physics. By focusing on the mechanism of measurements, Szilard made the first tentative steps toward a connection between quantum theory and thermodynamics—a topic that has just recently returned to the forefront of research. The realization is growing that interventions in nature at the most basic level can only be understood with the help of a delicate blend of quantum mechanics, which describes atomic systems, and thermodynamics, which rules the macroscopic world of the laboratory. In 1929, the Demon sacrificed the second of his many lives for this insight.

1950: The Demon (Age Eighty-three) Is Crushed by the Cost of Information Acquisition

The French physicist Leon Brillouin, internationally recognized for his pioneering work on the theory of crystals, had moved to the United States and joined IBM, where he became interested in the fundamental principles of computer science. He put them to use in a paper entitled "Maxwell's Demon Cannot Operate: Information and Entropy." Almost simultaneously, Dennis Gabor, the inventor of holography, and Norbert Wiener, the founder of the science of cybernetics, reached similar conclusions.

Instead of focusing on the act of measurement, as Szilard had done, Brillouin concentrated on the results of measurements, which he appropriately called information. While *descriptions* of measurement necessarily refer to specific pieces of apparatus, and are therefore complex and wordy, the *outcomes* of measurements are invariably numerical and brief. Thus the explanation of how a surveyor would go about measuring the size of my lot is a complicated story of theodolites and compass readings, but the final result, say 1.18 acres, is a concise nugget of information. Brillouin proceeded from the hope that if the analysis of the Demon could cast directly in terms of the information he collects and uses, without getting bogged down in descriptions of the way in which he gathers that information, the analysis would gain generality and power.

The impetus for examining Szilard's reasoning in terms of the concept of information was the birth of a new science called "information theory." In a landmark paper of 1948, Claude Shannon, a mathematician at the Bell Telephone Laboratories, initiated a program of research that would eventually become an elegant, rigorous, and powerful theory of communications.

An essential tool of this theory is a quantity for measuring the amount of information conveyed by a message. Suppose a message is encoded into some long number. To quantify the information content of this message, Shannon proposed to count the number of its digits. According to this criterion, 3.14159, for example, conveys twice as much information as 3.14, and six times as much as 3. Struck by the similarity between this recipe and the famous equation on Boltzmann's tomb (entropy is the number of digits of probability), Shannon called his formula the "information entropy."

At first glance this idea appears as paradoxical as Einstein's connection of energy with inertia. After all, information is desirable, useful, and valuable, whereas entropy connotes the opposite—disorder, dissipation, and waste. But the contradiction is only apparent, and is easily resolved by the addition of an adjective. Boltzmann's entropy, it turns out, is "missing information." For suppose we knew the exact velocity and position of every molecule in a quart of air. Our information would be maximal, order would be perfect, entropy would be at its very minimum. But of course we don't have this information: It is missing. When we count all the possible combinations of values of the speeds and positions that the air molecules *could* take on, we get *M*, the monster number estimated by David Ruelle. The letter *M* takes on a new significance: In addition to "monster" it stands for "missing." The number of its digits is the entropy.

In summary, entropy and information are closely related. The acquisition of information corresponds to a decrease in entropy. The more we know, the more order we impose on our universe. Conversely, the loss of information inherent in dissipative processes like the shuffling of a deck of cards, the diffusion of a drop of ink in a glass of water, or the shattering of a pane of glass entail an increase in entropy. Boltzmann's formula measures not only disorder, but also missing information.

Brillouin used Shannon's identification of information with entropy to exorcise the Demon. The sorting Demon decreases the entropy of a gas by his tricks, but in order to do so, he must first collect information about the molecules he watches. This activity, in turn, has a thermodynamic cost that can be calculated by means of Shannon's formula. Thus it turns out that by merely watching and measuring, the Demon raises the entropy of the world by an amount that saves the second law.

Thus Brillouin killed the Demon more decisively than Szilard. This time, his death led to the astonishing proposal that information is not just an abstract, impalpable, subjective construct of the mind, but a real, physical commodity with as much concreteness as work, heat, and energy. This notion was a radical departure from conventional thinking, and would henceforth reverberate through all speculations about the foundations of physics.

The linking of entropy with information by Brillouin, Wiener, and Gabor brought about the Demon's third demise. Though incorrect in detail, this radical proposal nevertheless ushered in a new era in physics. A hundred years earlier, the idea of an abstract, immaterial quantity called energy had suddenly cropped up and quickly established itself alongside Newton's intuitively more accessible conceptions of matter, motion, and force. "Information" is even more nebulous and subjective than energy, so its introduction into physics will not proceed as rapidly. But eventually, when physics makes contact with computer science and cognitive science—whose goal is nothing less than to understand human thought—the concept of information is sure to come into its own. And it was the Demon who made the first contact—right at the mid-century mark.

1982: The Demon (Age 115) Drowns in Garbage

The verdict handed down by Szilard and Brillouin—that the Demon cannot operate if measurements and information acquisition are properly accounted for—had been quietly accepted by a generation of physicists. Meanwhile, however, a growing number of mathematicians and computer scientists began to question its logic. Chief among them was Charles Bennett, a brilliant thinker at IBM whom the Nobel laureate Murray Gell-Mann has compared to ". . . a twelfth-century troubadour traveling from court to court in what is now the south of France. Instead of courtly love, Charlie sings of complexity and entropy, of quantum computers and quantum encipherment."

Drawing on the previous insights of his equally free-spirited and iconoclastic IBM colleague Rolf Landauer, Bennett took aim at the common belief that information acquisition, which includes both measurements and computations, is necessarily associated with irreversible losses of energy. He demonstrated that contrary to Brillouin's assertion, computers can in principle operate without losses. To this end he invented a series of imaginary mechanical computers that operate somewhat like well-oiled and exquisitely machined abacuses, whose friction can be reduced to an arbitrarily low level. Bennett enjoyed a decisive logical advantage over Brillouin. All he had to do to prove his point was to display *one, single* reversible computer, whereas Brillouin was only able to show that *some* devices involve irreversible losses, not that all of them do.

From the reversible computer, Bennett proceeded to the Demon. He argued that if the Demon can compute reversibly, he can also make measurements reversibly. The trick turned out to involve replacing the optical measuring devices favored by Brillouin and Gabor, who both happened to be experts in the theory of how light is created and absorbed, with mechanical machines that are much less sophisticated but can be made frictionless. In this way Bennett found an unexpected flaw in Brillouin's reasoning. Information, it turned out, *can* be collected and manipulated in a reversible way, entropy need not increase whenever measurements are performed, and the intelligent Demon is alive again.

Reading the papers of Landauer and Bennett is a strange experience. The Demon's flashlight and optical detectors give way to highly polished pistons and frictionless pressure gauges. Electronic devices and digital computers are replaced by strange contraptions that look like the Victorian calculators invented by Charles Babbage. Mechanics, the first branch of physics, takes over from electronics, its latest. It is amusing as well as reassuring to discover that old-fashioned machines, of the kind Sadi Carnot's father described, can still teach us lessons that are lost in the intricacies of their modern replacements.

Driven by the physicists' ingrained suspicion of the Demon, Bennett pursued his analysis to its final step—and discovered yet another surprise. Computation, he realized, requires temporary storage of information, whether it is on a ribbon of paper, an electronic memory, or a magnetic tape. According to Landauer the *destruction* of this information by erasure, by clearing a register, or by resetting the memory is irreversible. If the Demon commanded an infinite memory, or an infinite store of blank paper, he could indeed violate the second law. But since he doesn't—the universe is finite, after all—since he must periodically clear his notebooks and magnetic tapes for further computation, he will dissipate enough heat to save the second law.

In the end, Bennett's vision of the Demon was this: As

he has for a century, the goblin squats at his trapdoor between two gas-filled boxes, watching, measuring, figuring. But instead of processing all his observations in his head, he works with a little hand-driven calculator that prints its output on a ribbon of blank paper. As long as fresh ribbon is fed in from the outside, and used ribbon tumbles from the machine in an unending stream, the Demon can sort its molecules and draw on the energy contained in their random motion. If the paper ribbon is ignored in the analysis, the Demon violates the second law by extracting work from the gas without wasting any heat. (In practice it might actually be more efficient if he took Count Rumford's advice and just burned the paper: An infinite ribbon can supply an infinite amount of energy.)

However, if the Demon decides to work with a short ribbon that he recycles by periodic erasure, he will dissipate enough energy, generate enough entropy, and create enough disorder to save the validity of the second law. When the price for erasing useless, leftover numbers in the Demon's memory—of garbage disposal—is properly accounted for, the Demon cannot operate. He is thrown out of the house of science.

The lesson of this latest defenestration constitutes the most significant advance in our understanding of Maxwell's Demon since 1914. Neglecting the real cost of garbage disposal can fool us into believing that we live in a world of unlimited resources. On the other hand, accounting for it correctly may fundamentally change our economic forecasts. Bennett's analysis of the Demon demonstrates that information costs more in disposal than in acquisition, and reflects the world's belated realization that waste disposal must become an integral part of our economic reckoning.

Four times in this century the Demon has been thrown out of the window and pronounced dead. The first three times, he scrambled back, but for the time being the consensus of scientific opinion is against him. "Whatever his latest disguise may be," it is said, "he cannot violate the second law of thermodynamics." But that pronouncement doesn't seem to deter the restless sprite. Even if he cannot carry out the job Maxwell gave him, there are many other ways in which he can puzzle, annoy, trick, mislead, inform, and inspire the community of physicists. Although he is dead, his principal biographers feel that "this fanciful character seems more vibrant than ever."

19

THE NEW ENTROPY

The entropy concept . . . in spite of its age . . . has kept an untarnished lustre of novelty, an aura of unplumbed depth. It may well be that it holds further surprises in store for us.

—PETER LANDSBERG

To learn about the Demon's latest adventures, I went to visit Wojciech (Wojtek) Żurek at the Los Alamos National Laboratory in New Mexico. From the airport at Albuquerque, the road to Los Alamos leads past reclusive millionaires' estates nestled in the wooded hills around Santa Fe, and then through an arid valley dotted with gaudy gambling casinos built by Native Americans to lure the millionaires. After crossing the muddy Rio Grande, the road climbs up the steep side of a mesa, twists alarmingly a couple of times, and finally scrambles over the top to a breathtaking vista across the valley. The tortuous drive helped to explain why half a century ago the builders of the first atomic bomb chose this remote summit to go about their secret business.

After World War II, the Los Alamos National Laboratory branched out into a variety of scientific endeavors of a more peaceful nature, but weapons research continues. In fact, after Żurek's secretary phoned to make sure that I am an American citizen, I was expecting guards, gates, name tags, and all the other paraphernalia of national security, but when I found the quarters of the theoretical astrophysics group, an unprepossessing office building of government issue, its front door was wide open. A directory led me down the hall to a door bearing my host's name, along with assorted posters announcing conferences in farflung places in which he had participated as organizer or invited speaker—the badges of honor of the successful academic scientist. Żurek has an international reputation for his thoughtful and often highly original contributions to a number of problems ranging from the interpretation of quantum mechanics to the foundations of thermodynamics. What had brought his name to my attention in the first place was his computer simulation of the mechanical Demon.

Answering my knock, Żurek was tall, fortyish, dressed in a loose yellow shirt and the obligatory Western jeans, and grinning from ear to ear. "Welcome!" he said, and pumped my hand with vigor. A tangle of reddish-blond hair formed an immense aureole around his head and contrasted strikingly with a neatly trimmed Vandyke beard. His eyes, in deep sockets above a great beak of an aquiline nose, regarded me with warmth and interest. He reminded me strangely of the Demon of my imagination.

After I explained the purpose of my visit, Żurek, whose impeccable English has the characteristic inflection of the native speaker of Polish, proceeded to tell me about his idea for a new definition of entropy.

"What's wrong with the old one?" I asked.

Two things, he said. First, there is the shuffling trick, and second, after all these years the corpse of Maxwell's Demon is still twitching and needs to be attended to. With that I knew that I had come to the right mountain.

The shuffling trick refers to the fellow who shows you a deck of cards ordered by value and suit, shuffles it thoroughly, lets you verify that it is in complete disorder, shuffles it again, and then triumphantly lays it out on the table.

Expecting some miraculous restoration to order, you are surprised to find that the sequence of cards seems as random as it was before. The shuffler, however, smiles proudly and challenges you: "I'll bet *you* can't do that!"

Don't take the bet. The exact sequence on the table—whatever it turns out to be—is just as difficult for you to reproduce as any orderly pattern of the type you had expected.

This silly trick illustrates an important property of order. As soon as a sequence of cards is known—by being laid out on the table—it is in perfect order. It is no less orderly, for example, than a systematic sorting by suit and value. Regardless of its apparent randomness or obvious orderliness, the probability of a chance drawing of a specific, preannounced sequence from a shuffled deck is the same, and equals one in eight followed by sixty-seven zeros. Order, in other words, is difficult to define in an absolute way—like the concept of energy or of time. It is a very slippery term.

Applied to physics, the shuffling trick echoes the profound disjunction between the certainty of mechanics and the probabilistic nature of thermodynamics that had frustrated Maxwell and led him to turn to the Demon for help. Consider a bottle full of air, and assume that you know the position and velocity of every molecule. Then, in the context of classical mechanics, you know everything there is to know about the air and can predict exactly what will happen. For example, you can readily compute the energy stored in the bottle in the form of molecular motion, and you can confidently predict that the first law of thermodynamics, the principle of energy conservation, will hold. But the entropy of the configuration is zero! According to the conventional interpretation, there is no missing information, no randomness, no disorder, no entropy—just as in the case of the cards on the table. Hence, without invoking probabilities, you cannot compute the correct entropy of the air, nor, in consequence, formulate the second law. The second law seems to be incompatible with the normal procedures of classical mechanics.

Żurek's proposal to amend the definition of entropy is intended to move it away from the uncertainties and ambiguities associated with the calculus of chance and toward the definiteness that is characteristic of mechanical models of nature, and that Maxwell was accustomed to from his Devil-On-Two-Sticks.

As for the twitching Demon, Bennett's discovery that irreversibility resides in information disposal has put him into a peculiar position: As long as he keeps his used tape, he can temporarily violate the second law. It is only when he decides to erase discarded information in order to prepare for fresh records that he is able to set things right again. So whether or not the second law holds seems to depend on the whims of that obstreperous goblin. Surely the supreme law of nature should be formulated in a more robust fashion!

Żurek's new definition of entropy, which is designed to remedy this ambiguity, makes use of algorithmic randomness (aka algorithmic complexity)—a mathematical concept that was introduced by Russian and American mathematicians during the 1960s for the purpose of measuring the degree of randomness of a number, or any collection of data, without recourse to probability. For example, it serves to assess the relative randomnesses of the sequences 012345678901 . . . , 3141592653 . . . , and 417308629406. . . . Intuitively one feels that the first is less random—more orderly—than the other two because it is predictable. But a closer look reveals that the second one happens to be the beginning of the number pi with the decimal point left out—a universal constant whose digits are predictable out to infinity. So how do the three sequences stack up with respect to randomness?

Algorithmic randomness is defined as the length of the shortest computer program that can generate the number.

(The adjective "shortest" is necessary to avoid lengthening the program needlessly by the addition of redundant steps.) Fortunately it turns out that neither the exact form, nor the definition of the word "length," matter very much. In practice the way to measure length is to translate the computer program into a standard numerical form and then to count its digits. The program for the first example above might read, before translation into a number, like this: "List the natural numbers in ascending order from 0 to 9, then repeat until you reach the desired number of digits." The second one says: "Compute π by one of the standard formulas of calculus, omit the decimal point, and stop at the desired number of digits." The third one cannot be generated by a program because I just tapped it out at random on my laptop. The only recourse for reproducing my number is to copy it, so its only "program" is the sequence itself, and its algorithmic randomness is measured by the number of its digits. Randomness thus turns out to be almost zero for the first sequence, fairly small for the second one, because formulas for π are simple, and maximal for the third. Furthermore, if the three sequences were continued to infinity, the randomnesses of the first two would remain unchanged, while the value for the third one would become infinite. Likewise, the algorithmic randomness of a deck of cards begins at a minimum when there is a simple, predictable pattern and reaches its maximum when there is no order.

Żurek realized that since algorithmic randomness measures disorder even for a completely known deck of cards or bottle full of air, and since it manages to do this without reference to statistics or probability, it might play a useful role in a new definition of entropy.

Over lunch at the laboratory cafeteria, Żurek emphasized that his new entropy is purposely contrived to be numerically close to the old one, at least in most cases. "It had better be," he laughed, "or else one hundred and fifty years of theoretical and practical successes of the definition by Clausius and Boltzmann would blow my proposal out of the water!" But, he explained, the new definition of entropy is *conceptually* very different from the old one, and for that reason may prove to be controversial for a while.

"Will it replace the standard entropy?" I asked. Żurek smiled and shrugged. "I hope so," he admitted, but blamed himself for not doing enough to popularize it. I reminded him that it took Clausius about two decades to install *his* definition of entropy in the vocabulary of theoretical physics.

Żurek's principal innovation is to define entropy not from the perspective of some superior, all-knowing being who stands above and outside the system in question—in other words, the physicist—but "from the Demon's point of view." This phrase was to recur so often that afternoon that I began to suspect that the Demon has a special appeal for the boyish iconoclast in Żurek.

In Żurek's view, the Demon sits in his box and watches molecules. As far as he is concerned, there are only two kinds: Those he has measured and recorded, and those he has not. In his quest for control and orderliness, he notes two distinct types of disagreeable disorder, or entropy. The first, overwhelming type derives from his own ignorance. It is the entropy associated with the molecules he hasn't observed yet—Boltzmann's entropy of missing information. In addition there is another tiny, previously overlooked kind of disorder. In magnitude it differs from Boltzmann's entropy as much as a thimbleful of water differs from a lake. The information the Demon has already collected and recorded on his tape also contains disorder, and this is measured most appropriately by the tape's algorithmic randomness. Thus Żurek arrives at his definition of what he calls physical entropy: *Physical entropy is the entropy of missing information, plus the algorithmic randomness of the information that has been recorded.*

The cogency of the new concept is striking. Disorder, which is to say entropy, lurks in information, and information is divided into two kinds—that which is missing and that which is at hand. Both contribute to physical entropy.

With his new definition, Żurek has overcome the two difficulties he mentioned in the beginning. Entropy is defined absolutely and without reference to probability. Furthermore, if the Demon decided to erase some tape, entropy would move from the "known information" column into the "missing information" column without changing its sum total. The effect would be analogous to pouring water from a thimble into a lake without spilling a drop. Physical entropy, according to its new definition, does not depend on the Demon's decision to erase or to keep information: Its meaning is more robust than that of the conventional entropy.

I asked Żurek to explain how his entropy fits into the formulation of the second law. If there is no Demon, he began, entropy is entirely of the missing information kind, so the old theory applies unchanged. At the other extreme, if an omniscient Demon knows everything there is to know about a system, the second law also holds. There is no missing information, so the only entropy is that which is associated with the Demon's records. A simple, very orderly initial state is easy to describe, and thus has low algorithmic randomness, whereas a more disorderly state, which requires more words for its description, implies a higher randomness. Nature, like Ehrenfest's urns, proceeds from the simple to the complex, from the orderly to the disorderly, from low entropy to high entropy.

With or without an omniscient Demon, physical entropy increases naturally until equilibrium is reached. Once these two extreme cases are covered, it is not hard to believe that the intermediate case of a partially informed Demon can also be incorporated in the theory. In summary, the second law can readily be formulated in terms of

Żurek's concept of physical entropy. Careful analysis also reveals that, as expected, the second law prevents the Demon from pushing heat uphill from a cold box into a warmer one.

Although the new interpretation represents a sharpening of the language, it does not constitute a proof of the second law. It lifts the law off its traditional bed of probability theory and places it instead upon a futuristic pedestal of complexity theory, but whether this brings the proof any closer remains to be seen.

Żurek warned me that algorithmic randomness, which is central to the new entropy, is itself controversial. It is well and good to speak of "the shortest computer program that will reproduce a data set," but there is nothing in the definition to help you write that program. A person who has never heard of pi would assign an infinite complexity to the string of seemingly random digits that make up its numerical value, whereas a more sophisticated mathematician would assign it a low complexity. Who is right?

For describing a vessel full of gas the ambiguity inherent in the phrase "the shortest computer program" is not a serious problem. A listing of the positions and velocities of all the molecules is fairly straightforward, and can be made without the mathematical sophistication required to recognize the constant pi.

Nevertheless, the apparent subjectivity of algorithmic randomness presents problems of interpretation that will have to be resolved before Żurek's definition of entropy replaces the old one in common usage.

I wondered about applications. "How will this affect the engineers and chemists, who use entropy as their working tool? When will they have to take into account the change you're proposing?" Żurek laughed again. The same circumstance that protects his theory against criticism keeps it from having any immediate discernible consequences: The entropy added to Clausius's old formula by taking into account algorithmic complexity is exceedingly small in most cases, and negligible in practical applications. Nature sees to that by its choice of constants.

Like all subfields of physics, thermodynamics is characterized by certain constants whose values we measure but otherwise take for granted. Thus, special relativity is built upon the speed of light *c*, which happens to be very large. As long as the speeds of objects such as cars and apples don't come near *c*, the effects of special relativity will be immeasurably small. For that reason, Einstein's theory took a long time to be verified experimentally—even though its logical consistency had quickly become apparent. The opposite happened in quantum theory. The relevant number, Planck's quantum of action *h*, is so unimaginably small that its effects are difficult to discern. If *c* were smaller, and *h* larger, relativity and quantum mechanics would shape our everyday lives and intuitions as powerfully as does *g*, the acceleration of gravity.

The natural constant that takes the place of *c* and *h* in thermodynamics is Boltzmann's constant *k*, which converts absolute temperature into molecular energy, and the logarithm of probability into entropy. According to the theories of Brillouin, Wiener, and Gabor in the 1950s, *k* also relates information to entropy. Now, it so happens that *k* is minuscule. Measured in the scientific units for energy over temperature (joules per kelvin), *k* amounts to about ten trillionths of a trillionth—a number that is far beyond the pale of human understanding. The only reason *k* has any discernible effects at all is that when applied to the conventional entropy, it is multiplied by the number of molecules. Since that number, in turn, is usually measured in the trillions of trillions, it compensates for the insignificance of *k*. But in the case of Żurek's entropy of algorithmic complexity, such compensation is lacking. The quantity that multiplies *k* in this case is only the number of digits of information the Demon has managed to collect—

yielding a very small number indeed, and therefore a virtually negligible addition to the old entropy.

So why even bother with it? Because, Żurek bristled, it's a matter of principle. Like all scientific ideas, the concept of entropy, useful as it is, needs to be refurbished and updated and adjusted to new insights. Someday, when nanotechnology begins to manipulate much smaller numbers of molecules than we are used to, and when computers can store much vaster quantities of information than today, the two types of entropy will begin to approach each other in value, and the new theory will become amenable to experimental verification. In the meantime, it must continue to rely on its internal logic alone.

Late in the afternoon, weary from the cascade of ideas and calculations Żurek had poured over me, I asked him whether he thought the Demon was really finally dead. "Not at all" was the surprising answer, "he has just been underemployed all these years. Instead of sitting in a gas at equilibrium, foolishly trying to beat the second law—which he cannot do—he should be in some system that is far from equilibrium. And there he should use his massive intelligence to figure out ways of extracting energy in the most efficient way possible. He should apply the second law to some useful purpose, not try to subvert it!"

I thanked Żurek for giving me so freely of his time, and took my leave. It seemed to me that the Demon had finally met his match. For a century and a quarter, people have been afraid and suspicious of him, and have concentrated their efforts on trying to assassinate him. Here, at last, was someone who truly understood him, and tried to see things from his point of view. And then, instead of threatening to throw the poor little goblin out the window, he calmly ordered him to drop the futile task he had been given, shooed him out of his gloomy box, and gave him a fresh purpose in life.

20

FANTASTIC VOYAGE: THE DEMON ENTERS THE HUMAN BODY

Once we learn to expect theories to collapse and to be supplanted by more useful generalizations, the collapsing theory becomes not the gray remnant of a broken today, but the herald of a new and brighter tomorrow.

—ISAAC ASIMOV

James Clerk Maxwell invented the Demon as a means of harnessing the immense supply of energy stored in the incessant jiggling of molecules. This random, chaotic motion is often called "thermal noise" because when it is translated into sound by sensitive microphones, it comes out as a monotonous, steady hiss—a useless, grating noise. In its most primitive guise, the Demon was supposed to be a device for taming thermal noise by herding molecular motion in a particular direction and channeling its energy—a valve, as Maxwell called it; a ratchet, in Feynman's analysis; or a rectifier, in the terminology of electronics. Since 1912, physicists have realized with increasing conviction that such a device, which would violate the second law of thermodynamics, cannot work.

Rejected as a simple, mechanical machine, the Demon then came back endowed with the abilities to observe and to think, and he thereby helped to focus attention on those enigmatic commodities called "information" and "intelligence." Indeed, Maxwell's Demon has helped clarify their roles in physical science to some extent, but it remains for future research to raise their precision and power to the level of such universal conceptual tools as energy and entropy. In the meantime the Demon, back in his old, mechanical incarnation, has finally abandoned his futile quest and lowered his sights. As Wojciech Żurek recommended, he has left his cozy perch in a box full of warm gas and installed himself in more exposed, less tranquil quarters. The move has resulted in a spectacular rebirth; at the same time it has allowed the Demon to sneak into yet another field of research—this time the burgeoning science of molecular biology. At the venerable age of a hundred and thirty, Maxwell's minion has found another promising niche.

His latest reincarnation is an example of the time-honored precept that if you can't lick 'em, you should join 'em. If the mechanical Demon cannot extract energy from thermal noise, what is the simplest device that can? What modifications must be made to a valve or a ratchet that converts it into an engine for making use of thermal noise, all the while obeying the laws of thermodynamics? How can the Demon be brought to life? More exciting than the question, which is admittedly somewhat contrived, is the discovery that the answer may turn out to reveal how macromolecules manage to defy gravity and drift uphill in the human body and how muscles convert chemical energy into motion. Once more the cunning Demon has found his way into the center of scientific excitement.

Recent theoretical and experimental papers on the rectification of thermal noise invariably begin with a reference to Richard Feynman's ratchet machine, which anchors the discussion and establishes an intuitively appealing language. Recall the setup: A miniature windmill, unwitting descendant of the wheels of Villard de Honnecourt, Isaac Newton, Sadi Carnot, and Robert Mayer, is installed in a box full of warm gas, so that its vanes are randomly driven forward and backward by the incessant bombardment of molecules. In order to tame the motion, the windmill's shaft is equipped with a ratchet-and-pawl mechanism similar to those used in watches and a certain type of screwdriver. When the windmill turns forward, the pawl slides easily along one of the long, gentle slopes of the ratchet wheel and snaps down to the next one at each sharp, downward step—clicking along like the return twist of a ratchet screwdriver. But as soon as the shaft begins to turn backward, the pawl strikes one of the short, steep steps, gets stuck, and prevents further motion. The result is the conversion of random forward and backward jiggling into pure forward motion.

Feynman conceded that when his device first starts out, it will extract energy from the gas. It might even lift a little weight hung from a string wrapped around the shaft. However, friction between the metal parts of the mechanism soon heats up the pawl to the point where it begins to bounce up and down uncontrollably, like a drop of water on a hot stove. It starts to neglect its job and occasionally allows the shaft to turn in the wrong direction. Feynman showed that as long as the entire system is at the same temperature, or, as physicists say, in thermal equilibrium, the motion of the windmill averages out to zero. This discussion, which formed part of a memorable lecture entitled "The Distinction of Past and Future" given at Cornell University in 1964, and was published in *The Character of Physical Law* as well as the celebrated *Feynman Lectures*, has served a generation of physicists as the most persuasive indictment of the mechanical Demon. The time has come to improve its design.

The direction of the next step was indicated by Feynman himself. He pointed out that it is easy to make the thing work: Just keep the ratchet cool. If there is a convenient reservoir nearby to absorb the heat created by friction, and thereby to subdue the Brownian motion of the pawl, the whole machine will operate like a conventional engine, which takes heat from the gas in the box, converts some of it into work, and discards the rest into a reservoir. As a matter of fact, Feynman even claimed that the engine can be reversed: If the pawl is heated by some external means and maintained at a higher temperature than the gas in the box, it will push down on the long, gentle slopes of the ratchet hard enough to move them backward before it hops up onto the next step. Thus, Feynman claims, it will force the windmill to turn backward. Although this scenario is a bit difficult to imagine, and highly impractical to check out, we would be wise to give Feynman the benefit of the doubt and concede the point.

Ultimately Feynman concluded that in equilibrium, thermal noise cannot cause motion or do work, even in the presence of a ratchet or similar element. However, when a temperature difference is introduced by means of an external heat reservoir, which can function either as a source or a sink, the device *can* extract energy from thermal noise.

With that remark, Feynman's little motor became much more than a textbook illustration. If it could actually be made to work, it would serve as a prototype for a machine that harnesses nature's ubiquitous Brownian motion for useful purposes. The greatest interest would be excited among physiologists, who struggle to explain how molecules and larger organisms manage to move around the human body unaided by mechanical pumps and motors. A scheme for exploiting Brownian motion—with the help of body heat—might just do the trick.

Unfortunately, it doesn't. R. Dean Astumian, a physician like Mayer, as well as professor of biochemistry and molecular biology at the University of Chicago, has examined the uses of thermal noise by the body, and found that while the suggestion works in principle, the actual numbers don't work out. The problem is that the requisite temperature differences between cold and hot reservoirs just don't exist in the tiny confines of biological environments.

It turns out that if you can't maintain a large temperature difference between the two sides of a cell, say, then you can't use it to transport a molecule across the cell.

Undeterred by this pessimistic assessment, Astumian and other scientists in several laboratories around the world began to turn the investigation away from the cul-de-sac that Maxwell had inadvertently entered. The goal was to find out whether other influences, besides temperature differences, can make the ratchet work. And the answer turned out to be yes! The crucial additional ingredient was found in an unexpected element of the ratchet operation. Whereas Feynman had been very careful in his examination of the *geometry* of the ratchet, very little had been said about its timing. And here, it turned out, was the clue. If the machine was turned off at crucial intervals, in a regular cycle, it can be made to work.

In order to see how this comes about, we might equip Feynman's ratchet with a clock-driven motor for lifting the pawl at regular intervals, but the resulting Rube Goldberg machine would be too cumbersome to analyze with confidence. So let us banish Feynman's wheel into the attic of the history of science to join the dusty mechanical models of Victorian physics, and consider a simpler, sleeker, more elegant version of a ratchet. In fact, let us examine one that has actually been made to operate in the laboratory, unlike Feynman's imaginary windmill.

In 1995, the French physicist Albert Libchaber, a consummate experimenter who fifteen years earlier had been among the first to confirm the startling predictions of the emerging theory of chaos, teamed up with colleagues at the École Normale Superieure in Paris to build what they call an "Optical Thermal Ratchet." Only the name remains of Feynman's quaint engine—the actual device is a marvel of modern technology.

A tiny plastic pellet, with a density closely matching that of water, floats submerged in a bath of microscopic proportions. Buffeted from all directions by water molecules, it jiggles with Brownian excitement. In order to control the ball, Libchaber and his team nudge it into a cleverly designed version of a set of "optical tweezers," a gossamer structure made not of metal or glass or plastic, but of laser light. Using nothing more substantial than the minuscule pressure exerted by light on translucent materials, they place the ball onto the optical equivalent of a circular ramp with four hills and four valleys—a luminous miniature roller coaster. The "ratchet" part of the system, which provides the necessary spatial asymmetry, is found in the profile of the hills, which have long, gentle ascents followed by short, steep descents.

Once the ball has settled into one of the valleys of the roller coaster, where it trembles with Brownian motion, nothing more happens: Thermal noise, even in the presence of unsymmetric spatial structures, does not result in motion.

But since the roller coaster is made of light, it is easy to modify. So every now and then, at regular time intervals and for specified periods, the hills are purposely switched off—leaving only a circular track. As soon as that happens, the ball, propelled by thermal noise, begins an excursion in one direction or another. Left to its own devices, it would err back and forth along the periphery of the circle, with no net displacement in the end. But it isn't left to itself, for soon the hills are switched back on. After they have reappeared, the ball is just as likely to find itself trapped in the valley immediately ahead of it as it is to roll back into the valley in which it started. In contrast, the probability that it has wandered all the way back, past the long slope it started on, to the valley behind is very small indeed. The net effect—marvelous to behold—is that the ball moves forward at a slow and halting pace but doesn't stop. The speed with which it moves depends on the spatial and temporal details of the blinking roller coaster, but the principle has been established: Thermal noise has been rectified. Brownian motion has been tamed. The ratchet works at last.

It is important to note at once that the second law is not in jeopardy. In fact, the ball receives more energy from the intermittent laser light than from the water molecules, so from the point of view of thermodynamics, the apparatus constitutes nothing but a rather inefficient motor. Nevertheless, the ball moves *because* of, not *in spite* of, thermal noise. Directed motion has been achieved without gravity, without macroscopic electrical or magnetic forces such as those that propel particles around accelerators, without large temperature differences like those found in boiling pots and the cylinders of combustion motors, and without a large difference in chemical composition characteristic of a battery. By learning to sculpt light, Libchaber and his team have succeeded in fashioning a little Demon whose activities they can control from afar. Their purpose was not so much to design a practical machine as to demonstrate the fundamental mechanism for taming thermal noise.

Now that the principles have been demonstrated, theoretical progress will follow rapidly. Mathematical physicists will invent equations, based on the laws of mechanics and thermodynamics, for modeling devices that incorporate thermal noise, a ratchet, and cyclical, or possibly random, variations in timing. The requisite asymmetry might even reside in the timing cycle instead of the spatial configuration—the possibilities are without limit. In any case, Maxwell's Demon will be reduced to a theoretical specter, and we will no longer be able to imagine him as a wizened homunculus. But what he will have lost in poetry he will have gained in power.

"The use of noise in technological applications is still in its infancy, and it is far from clear what the future holds," Astumian wrote in the spring of 1997. He ventured to predict, though, that the molecular motors and pumps of the twenty-first century will have more in common with chemical processes than with the mechanical gears, levers, springs, and cogs that compose conventional steam engines, windmills, and water turbines. Chemical processes resemble thermal ratchets in that they are driven by heat in a direction imposed by tiny molecular electrical forces. Astumian's observation implies that the impressive micrographs of molecule-size plumbing and machines that have begun to appear under the rubric of nanotechnology may be misleading. It may be that if we can learn from nature's subtle designs at the atomic level, we won't have to struggle to transpose our own cruder inventions from the macroscopic scale to the very small. Nor will the transformation of heat into useful motion necessarily continue to depend on the wasteful discrete power strokes of conventional motors, but will instead proceed smoothly and continuously in uncountable microscopic steps. If all this comes about, we will have to thank the Demon for showing the way.

More exciting than the promise of smoother nanomachinery is the possibility that the principles of the thermal ratchet underlie biochemical processes. In particular, Astumian has proposed ways in which spatial asymmetry and cyclical modulation, both necessary ingredients of Brownian motors, and both achieved by artificial means in the optical ratchet, could occur in nature. The former, he believes, might be provided by nothing more complex than a chain of long molecules, each with a positive charge at the front end and a corresponding negative charge at the rear, lined up head to tail like elephants in a parade. In this arrangement the positive/negative intervals, each as long as an entire molecule (elephant), alternate with very short negative/positive intervals across the interstices (between two elephants). To a passing charged atom, the chain presents an electrical sawtooth pattern, or a ratchet. The other ingredient of a thermal ratchet, a cyclical variation in time, may be provided by oscillating reactions of chemicals far from equilibrium.

In this way Astumian has assembled, at least hypothetically, the tools needed to construct a natural motor for moving chemicals and particles through living tissue: thermal noise, a ratchet, and a timing mechanism. Is this how the body pumps vital nutrients across tissue walls? Is this how muscles contract? Is this how the body works? If it turns out that way, we will be able to invert Robert Mayer's insight—that the human body imitates a steam engine—and design new engines that imitate the body.

This fusion of physics, chemistry, biology, and engineering will be a triumph of the old science of thermodynamics, and a worthy monument to the memory of Maxwell's Demon.

NOTES

Introduction

PAGE

xi "Imagination is more important than knowledge": Bergen Evans, *Dictionary of Quotations* (New York: Delacorte Press, 1968), p. 340.

xvii "very observant and neat-fingered being": Harvey S. Leff and Andrew F. Rex, *Maxwell's Demon* (Princeton: Princeton University Press, 1990), p. 5. (Hereafter this book, which is my principal and indispensable source on the Demon, will be cited as MD.)

xviii "Maxwell's intelligent demon": MD, p. 34.

"The word 'demon' ": MD, p. 5.

Socrates' demon: Robert Nicol Cross, *Socrates—the Man and His Mission* (Freeport, N.Y.: Books for Libraries Press, 1970), p. 250.

xix "I have now conducted you": Quoted by Johannes Lohne in "Thomas Harriott (1560–1621): The Tycho Brahe of Optics," *Centaurus*, vol. 6, no. 2 (1959), p. 115.

"restless and lovable poltergeist": MD, p. 16.

two hundred references: MD, p. 331.

xx "Useful as it is": John A. Wheeler, "Law without Law," in John A. Wheeler and Wojciech H. Żurek, eds., *Quantum Theory and Measurement* (Princeton: Princeton University Press, 1983), p. 194.

Chapter 1

PAGE

3 "The laws of thermodynamics smell of their human origin": I have condensed the original statement by Percy

W. Bridgman in *The Nature of Thermodynamics* (Gloucester, Mass.: Peter Smith, 1969), p. 3, which reads: "It must be admitted, I think, that the laws of thermodynamics have a different feel from most of the other laws of the physicist. There is something more palpably verbal about them—they smell more of their human origin."

"I was led into these investigations": Sanborn C. Brown, *Benjamin Thompson, Count Rumford* (Cambridge, Mass.: MIT Press, 1981), p. 195. (This is my principal source on Rumford and is hereafter cited as Brown.)

4 a force of several tons: W. J. Sparrow, *Knight of the White Eagle: A Biography of Benjamin Thompson, Count Rumford* (London: Hutchison, 1964), p. 218.

5 "at 2 hours and 30 minutes": Count Rumford, "Inquiry Concerning the Source of Heat Excited by Friction," *Philosophical Transactions* (London), vol. 88 (1798), p. 80, reprinted in *Collected Works of Count Rumford*, Sanborn C. Brown, ed. (Cambridge, Mass.: Harvard University Press, 1968), vol. 1, p. 3.

7 "these computations shew": Brown, p. 195.

"abstruse speculation": Brown, p. 200.

The list began with caloric: William H. Brock, *The Norton History of Chemistry* (New York: W. W. Norton, 1993), p. 118.

positive and negative electricity: Ibid., p. 153.

9 twelfth century: Eugene Hecht, *Physics in Perspective* (Reading, Mass.: Addison-Wesley, 1980), p. 180.

"heat itself, its essence and quiddity": Sparrow, p. 215.

"heat being nothing else": Stephen G. Brush, *Kinetic Theory* (New York: Pergamon, 1965), vol. I, p. 6, footnote.

11 "I have often": Brown, p. 195.

Chapter 2

PAGE

13 "[Thermodynamics] is the only physical theory": Albert Einstein, "Autobiographical Notes," in *Albert Einstein: Philosopher-Scientist*, Paul Schillp, ed. (Evanston, Ill.: The Library of Living Philosophers, 1949), p. 32.

14 "Many a time": Roland Bechmann, *Villard de Honnecourt* (Paris: Picard, 1991), p. 249.

15 Leonardo, an experienced engineer: Ibid., p. 251.

cold fusion: Gary Taubes, *Bad Science: The Short Life and Very Hard Times of Cold Fusion* (New York: Random House, 1993).

Chapter 3

PAGE

19 "It is important to realize": Richard Feynman, *The Feynman Lectures on Physics*, vol. I (Reading, Mass.: Addison-Wesley, 1963), p. 4.2.

Thomas Kuhn: Thomas Kuhn, *The Essential Tension* (Chicago: University of Chicago Press, 1977), p. 66.

20 "a few days after our arrival": Kenneth L. Caneva, *Robert Mayer and the Conservation of Energy* (Princeton: Princeton University Press, 1993), p. 3. (This is my source for Robert Mayer and the quotations in this chapter.)

Chapter 4

PAGE

28 "I should like to write": Heike Kammerling Onnes, quoted in *Dictionary of Scientific Biography*, vol. XII (New York: Scribner, 1976), p. 220.

29 probably apocryphal: Donald S. Cardwell, *James Joule—A Biography* (Manchester, U.K.: Manchester University

Press, 1989), p. 88. (This is my source for James Joule, unless otherwise noted.)

30 "I shall lose no time": Gerald Holton and Stephen G. Brush, *Introduction to Concepts and Theories in Physical Science,* 2nd edition (Princeton: Princeton University Press, 1985), p. 273.

32 scientists at Cambridge University: Heinz Otto Sibum, "Reworking the Mechanical Value of Heat: Instruments of Precision and Gestures of Accuracy in Early Victorian England," *Studies in History and Philosophy of Science,* vol. 26, no. 1 (1995), p. 73.

"and since constant practice": Ibid., fn 107.

Chapter 5

PAGE

35 "Nothing in the whole range": William Thomson, Lord Kelvin, quoted in Sadi Carnot, *Reflexions on the Motive Power of Fire,* Robert Fox, ed. (Manchester, U.K.: Manchester University Press, 1986), p. 1.

36 "You beastly First Consul": Quoted in Sadi Carnot, *Reflections on the Motive Power of Fire,* E. Mendoza, ed. (Gloucester, Mass.: Peter Smith, 1977), p. xi. (This is my principal source for Carnot and the quotations in this chapter.)

45 another more poignant reason: Jean-Pierre Maury, *Carnot et la Machine à Vapeur* (Paris: Presses Universitaires de France, 1986), p. 111.

Chapter 6

PAGE

47 "Once or twice": C. P. Snow, *The Two Cultures and a Second Look* (Cambridge, U.K.: Cambridge University Press, 1969), p. 14.

51 "I have not yet disclosed": Gerald Holton, *Thematic Ori-*

gins of Scientific Thought (Cambridge, Mass.: Harvard University Press, 1973), p. 51.

Chapter 7

PAGE

56 "The law that entropy always increases": Arthur Eddington, *The Nature of the Physical World* (New York: Macmillan, 1929), p. 74.

The physicist Gerald Holton: Gerald Holton, *Thematic Origins of Scientific Thought* (Cambridge, Mass.: Harvard University Press, 1973), p. 47.

57 "The noblest aim of all theory": Gerald Holton, *The Advancement of Science, and Its Burdens* (Cambridge, U.K.: Cambridge University Press, 1986), p. 15.

Theories, in other words: From M. J. Cohen and J. M. Cohen, *The Penguin Dictionary of Modern Quotations* (London: Penguin, 1971).

61 "since I think it is better": Rudolf Clausius, "The Second Law of Thermodynamics," in Jefferson Hane Weaver, *The World of Physics,* vol. I (New York: Simon & Schuster, 1987), p. 741.

Chapter 8

PAGE

63 "The universe is made of stories": From "The Speed of Darkness" in *A Muriel Rukeyser Reader,* Jan Heller Levi, ed. (New York: W. W. Norton, 1994), p. 231.

Chapter 9

PAGE

67 "Nature, it seems": Piet Hein, *Grooks,* quoted in Jefferson Hane Weaver, *The World of Physics,* vol. I (New York: Simon & Schuster, 1987), p. 623.

"I am never content": Alan L. Mackay, *A Dictionary of Scientific Quotations* (Bristol, U.K.: Adam Hilger, 1991), p. 239.

68 "the atomic *fact*": Richard Feynman, *The Feynman Lectures on Physics* vol. I, (Reading, Mass.: Addison-Wesley, 1963), pp. 1–2.

70 Bernoulli's model: Daniel Bernoulli, "Kinetic Theory of Gases," in Jefferson Hane Weaver, *The World of Physics*, vol. I (New York: Simon & Schuster, 1987), p. 597.

71 "The elasticity of air": ibid., p. 599.

Chapter 10

PAGE

76 "It is truth": Quoted by James R. Newman in *The World of Mathematics* (New York: Simon & Schuster, 1956), p. 1361.

"the miller": Richard Westfall, *Never at Rest* (Cambridge, U.K.: Cambridge University Press, 1980), p. 60.

he watched the needle turn: Abraham Pais, *Subtle Is the Lord* (Oxford, U.K.: Oxford University Press, 1982), p. 37.

77 "What's the go o' that?": Lewis Campbell and William Garnett, *The Life of James Clerk Maxwell* (London: Macmillan, 1882; reprint, with a selection of letters from the 2nd edition, New York: Johnson Reprint, 1969), p. 28. (This is my principal source on Maxwell's life.)

79 "Maxwell's equations": The modern succinct version of these equations was developed after Maxwell's death by Oliver Heaviside. Cf. Bruce J. Hunt, *The Maxwellians* (Ithaca, N.Y.: Cornell University Press, 1991), p. 125.

80 "I have been carried": James Clerk Maxwell, "Address to the Mathematical and Physical Sections of the British Association," in *The Scientific Papers of James Clerk Maxwell*, W. D. Niven ed., (New York: Dover, 1965), vol. II, p. 216.

Chapter 11

PAGE

85 "There are few laws": Quoted in *Change and Chance*, John Ogborn, ed. (Harmondsworth, U.K.: Penguin Books, 1972), p. 29.

"Logic is conversant": Lewis Campbell and William Garnett, *The Life of James Clerk Maxwell* (London, 1882; reprint, New York: Johnson Reprint, 1969), p. 143.

86 "Jolly little beggars": David Wilson, *Rutherford—Simple Genius* (Cambridge, Mass.: MIT Press, 1983), p. 114.

"Let us suppose": Stephen G. Brush, *The Kind of Motion We Call Heat* (Amsterdam: North-Holland, 1976), p. 161.

87 Paracelsus: William H. Brock, *The Norton History of Chemistry* (New York: W. W. Norton, 1993), p. 51.

Chapter 12

PAGE

92 "The conception of the 'sorting demon' is": MD, p. 291.

"very observant": MD, p. 5.

94 "This reduces the Demon": MD, p. 6.

96 "It must heat up": MD, p. 11.

97 As recently as 1992: P. A. Skordos and W. H. Żurek, "Maxwell's Demon, Rectifiers, and the Second Law: Computer Simulation of Smoluchowski's Trapdoor," in *American Journal of Physics*, vol. 60 (1992), p. 876.

"Such a device might": MD, p. 125.

Chapter 13

PAGE

99 "The future belongs to those": Quoted by Eric M. Rogers, *Physics for the Inquiring Mind* (Princeton: Princeton University Press, 1960), p. 395.

100 "On the eighth of June": Engelbert Broda, *Ludwig Boltzmann* (Woodbridge, Conn.: Ox Bow Press, 1983), p. 25. (This is my pricipal source on the life of Boltzmann.)
"The scientist asks not": Ludwig Boltzmann, *Theoretical Physics and Philosophical Problems*, (Boston, Mass.: Reidel, 1974), p. 13.

101 Eldorado: Ludwig Boltzmann, "A German Professor's Trip to Eldorado," abridged and translated by B. Schwarzschild, *Physics Today*, vol. 45, no. 1 (January 1992), p. 44.
"Father gets worse every day": Peter Coveney and Roger Highfield, *The Arrow of Time* (London: W. H. Allen, 1990), p. 175.

102 "the development of theory": Boltzmann, *Theoretical Physics and Philosophical Problems*, p. 33.

103 "At first the variations": Broda, p. 22.

106 "entropy equals the number of digits": This simplified formulation is due to David Ruelle, *Chance and Chaos* (Princeton: Princeton University Press, 1991).

Chapter 14

PAGE

114 "Nothing is destroyed": *Dictionary of Scientific Biography*, vol. XIII (New York: Scribner, 1976), p. 380.
"The earth shall wax old": Crosbie Smith and M. Norton Wise, *Energy and Empire: A Biographical Study of Lord Kelvin* (Cambridge, U.K.: Cambridge University Press, 1989), p. 331.
"Within a finite period": Stephen Brush, "Thermodynamics and History," *The Graduate Journal*, vol. 7 (1967), p. 494.

115 "I am never content . . . I often say": Alan L. Mackay, *A Dictionary of Scientific Quotations* (Bristol, U.K.: Adam Hilger, 1991), p. 239.

116 Thomson read it with distaste: Stephen G. Brush, "Thermodynamics and History," *The Graduate Journal* vol. 7

(1967), p. 477. (This is my primary source for the remainder of this chapter.)

Chapter 15

PAGE

122 "The visible world": Quoted in Art Hobson, *Physics: Concepts and Connections* (Englewood Cliffs, N.J.: Prentice Hall, 1995), p. 293.

124 "This argument is amusing": Albert Einstein, *The Collected Papers of Albert Einstein*, vol. 2, John Stachel, ed. (Princeton: Princeton University Press, 1989), p. 268.

Chapter 16

PAGE

129 "The essence of nowness": Quoted in *Time's Arrows Today*, Steven Savitt, ed., (Cambridge, U.K.: Cambridge University Press, 1995), p. 7.

"what I admired most": Quoted in Jeremy Bernstein, *Quantum Profiles* (Princeton: Princeton University Press, 1991), p. 164.

130 "The views of space and time": Hermann Minkowski, "Space and Time," in *The Principle of Relativity*, Arnold Sommerfeld, ed. (New York: Dover, 1923), p. 75.

132 He felt that there is something essential: Cf. Paul A. Schillp, ed., *The Philosophy of Rudolf Carnap* (La Salle, Ill.: Open Court, 1963), p. 37.

Chapter 17

PAGE

136 "Time flows on": MD, p. 93.

139 "Concerning Demons": MD, p. 5.

140 Ruelle reports: Ruelle, p. 100.

142 "the supreme Law of Nature": Arthur Eddington, *The Nature of the Physical World* (New York: Macmillan, 1929), p. 74.

143 "That the world started": Quoted in Jefferson Hane Weaver, *The World of Physics* vol. I (New York: Simon & Schuster, 1987), p. 772.

"The thermodynamic arrow of time": Murray Gell-Mann, *The Quark and the Jaguar* (London: Little, Brown, 1994), p. 227.

Chapter 18

PAGE

145 "Let us stop here": MD, p. 31.

147 These devices: Cf. L. C. Biedenharn and J. C. Solem, "A Quantum-Mechanical Treatment of Szilard's Engine," *Foundations of Physics*, vol. 25 (1995), p. 1221.

"Measurements themselves": MD, p. 124.

152 "a twelfth-century troubador": Murray Gell-Mann, *The Quark and the Jaguar* (London: Little, Brown, 1994), p. 100.

155 "this fanciful character": MD, p. 2.

Chapter 19

PAGE

156 "The entropy concept": Alan L. Mackay, *A Dictionary of Scientific Quotations* (Bristol, U.K.: Adam Hilger, 1991), p. 146.

157 a new definition of entropy: W. H. Żurek, "Algorithmic randomness and physical entropy," *Physical Review A*, vol. 40 (1989), p. 4731.

158 sixty-seven zeros: Martin Goldstein and Inge F. Goldstein, *The Refrigerator and the Universe: Understanding the Law of Energy* (Cambridge, Mass.: Harvard University Press, 1993), p. 406.

Chapter 20

PAGE

166 "Once we learn": Isaac Asimov quoted in Timothy Ferris, *The World Treasury of Physics, Astronomy, and Mathematics* (New York: Little, Brown, 1991), p. 783.

169 the actual numbers don't work out: R. Dean Astumian, "Thermodynamics and Kinetics of a Brownian Motor," *Science*, vol. 276 (1997), p. 917.

170 In 1995, the French physicist: L. P. Faucheux, L. S. Bourdieu, P. D. Kaplan, and A. J. Libchaber, "Optical Thermal Ratchet," *Physical Review Letters*, vol. 74 (1995), p. 1504.

172 The requisite asymmetry: Dante R. Chialvo and Mark M. Millonas, "Asymmetric Unbiased Fluctuations Are Sufficient for the Operation of a Correlation Ratchet," *Physics Letters A*, vol. 209 (1995), p. 26.

"The use of noise": Astumian, p. 922.

INDEX

ABOUT THE AUTHOR

HANS CHRISTIAN VON BAEYER, Chancellor Professor of Physics at the College of William and Mary in Williamsburg, Va., grew up in Germany, Switzerland, and Canada. After working on theoretical particle physics during the early part of his career, he now devotes his time to teaching and the popularization of science. With his wife, the art historian Barbara Watkinson, and their two daughters, he spends summers in Paris, where his work is called "vulgarisation de la science."

ABOUT THE TYPE

This book was set in Berling. Designed in 1951 by Karl Erik Forsberg for the Typefoundry Berlingska Stilgjuteri AB in Lund, Sweden, it was released the same year in foundry type by H. Berthold AG. A classic old-face design, its generous proportions and inclined serifs make it highly legible.